Review of EPA's Environmental Monitoring and Assessment Program: Overall Evaluation

Committee to Review the EPA's Environmental Monitoring and Assessment Program

Board on Environmental Studies and Toxicology

Water Science and Technology Board

Commission on Life Sciences

Commission on Geosciences, Environment, and Resources

National Research Council
Washington, D.C. 1995

NOTICE: The project that is the subject of this report was approved by the Governing Board of the National Research Council, whose members are drawn from the councils of the National Academy of Sciences, the National Academy of Engineering, and the Institute of Medicine. The members of the board responsible for the report were chosen for their special competences and with regard for appropriate balance.

This report has been reviewed by a group other than the authors according to procedures approved by a Report Review Committee consisting of members of the National Academy of Sciences, the National Academy of Engineering, and the Institute of Medicine.

Support for this project was provided by the U.S. Environmental Protection Agency under Agreement No. 68C00082/C.

Library of Congress Catalog Card No. 95-68895
International Standard Book Number 0-309-05286-6

Additional copies of this report are available from:

National Academy Press
2101 Constitution Avenue, N.W.
Box 285
Washington, D.C. 20055
800-624-6242
202-334-3313 (in the Washington Metropolitan Area)

B-550

Copyright 1995 by the National Academy of Sciences. All rights reserved.

Cover by John Eberhard, Pittsburgh, PA.

Printed in the United States of America

COMMITTEE TO REVIEW THE EPA'S ENVIRONMENTAL MONITORING AND ASSESSMENT PROGRAM

RICHARD FISHER, *Chair*, Texas A&M University, College Station
PATRICK L. BREZONIK, University of Minnesota, St. Paul
INGRID C. BURKE, Colorado State University, Ft. Collins
LOVEDAY L. CONQUEST, University of Washington, Seattle
THURMAN L. GROVE, North Carolina State University, Raleigh
JOHN E. HOBBIE, Marine Biological Laboratory Ecosystems Center, Woods Hole, Massachusetts
TIM K. KRATZ, University of Wisconsin, Madison
ANNE E. MCELROY, State University of New York, Stony Brook
JOHN PASTOR, University of Minnesota, Duluth
JAMES N. PITTS, JR., University of California, Irvine
SAUL SAILA, University of Rhode Island, Kingston
TERENCE R. SMITH, University of California, Santa Barbara
SUSAN STAFFORD, Oregon State University, Corvallis
MICHAEL J. WILEY, University of Michigan, Ann Arbor

Liaison from the Board on Environmental Studies and Toxicology

KRISTIN SHRADER-FRECHETTE, University of South Florida, Tampa

National Research Council Staff

SHEILA D. DAVID, Study Director, Water Science and Technology Board
DAVID J. POLICANSKY, Study Director, Board on Environmental Studies and Toxicology
ANITA A. HALL, Senior Project Assistant, Water Science and Technology Board
SHIRLEY F. JONES, Project Assistant, Board on Environmental Studies and Toxicology

BOARD ON ENVIRONMENTAL STUDIES AND TOXICOLOGY

PAUL G. RISSER, *Chair*, Miami University, Oxford, Ohio
MICHAEL J. BEAN, Environmental Defense Fund, Washington, D.C.
EULA BINGHAM, University of Cincinnati, Cincinnati, Ohio
EDWIN H. CLARK, Clean Sites, Inc., Alexandria, Virginia
ALLAN H. CONNEY, Rutgers University, New Jersey
ELLIS COWLING, North Carolina State University, Raleigh
JOHN L. EMMERSON, Eli Lilly & Company, Greenfield, Indiana
ROBERT C. FORNEY, Consultant, Unionville, Pennsylvania
ROBERT A. FROSCH, Harvard University, Cambridge, Massachusetts
KAI LEE, Williams College, Williamstown, Massachusetts
JANE LUBCHENCO, Oregon State University, Corvallis
GORDON ORIANS, University of Washington, Seattle, Washington
FRANK L. PARKER, Vanderbilt University, Nashville, Tennessee
GEOFFREY PLACE, Consultant, Hilton Head, South Carolina
DAVID P. RALL, Consultant, Washington, D.C.
LESLIE A. REAL, Indiana University, Bloomington, Indiana
KRISTIN SHRADER-FRECHETTE, University of South Florida, Tampa, Florida
BURTON H. SINGER, Princeton University, Princeton, New Jersey
MARGARET STRAND, Eckert, Seamans, Cherin & Mellott, Washington, D.C.
GERALD Van BELLE, University of Washington, Seattle, Washington
BAILUS WALKER, JR., University of Oklahoma, Oklahoma City

Staff

JAMES J. REISA, Director
DAVID J. POLICANSKY, Associate Director and Program Director for Applied Ecology and Natural Resources
CAROL MACZKA, Program Director for Toxicology and Risk Assessment Program

KULBIR BAKSHI, Program Director for Committee on Toxicology
LEE R. PAULSON, Program Director for Information Systems and Statistics
RAYMOND A. WASSEL, Program Director for Environmental Sciences and Engineering
BERNIDEAN WILLIAMS-SMITH, Administrative Associate

WATER SCIENCE AND TECHNOLOGY BOARD

DAVID L. FREYBERG, *Chair*, Stanford University, California
BRUCE E. RITTMANN, *Vice Chair*, Northwestern University, Evanston, Illinois
LINDA M. ABRIOLA, University of Michican, Ann Arbor
J. DAN ALLEN, Chevron U.S.A., Inc., New Orleans, Louisiana
PATRICK L. BREZONIK, University of Minnesota, St. Paul
WILLIAM M. EICHBAUM, The World Wildlife Fund, Washington, D.C.
WILFORD R. GARDNER, University of California, Berkeley
WILLIAM L. GRAF, Arizona State University, Tempe
THOMAS M. HELLMAN, Bristol-Myers Squibb Company, New York, New York
CHARLES C. JOHNSON, Jr., U.S. Public Health Service, Washington, D.C. (Retired)
CAROL A. JOHNSTON, University of Minnesota, Duluth
WILLIAM M. LEWIS, JR., University of Colorado, Boulder
CAROLYN H. OLSEN, Brown and Caldwell, Pleasant Hill, California
CHARLES R. O'MELIA, Johns Hopkins University, Baltimore, Maryland
IGNACIO RODRIGUEZ-ITURBE, Texas A&M University, College Station
HENRY J. VAUX, JR., University of California, Riverside

Staff

STEPHEN D. PARKER, Director
SHEILA D. DAVID, Senior Staff Officer
CHRIS ELFRING, Senior Staff Officer
GARY KRAUSS, Staff Officer
JACQUELINE MACDONALD, Senior Staff Officer
ETAN GUMERMAN, Research Associate
JEANNE AQUILINO, Administrative Specialist

ANITA A. HALL, Administrative Assistant
ANGELA BRUBAKER, Project Assistant
MARY BETH MORRIS, Senior Project Assistant
GREGORY NYCE, Senior Project Assistant

COMMISSION ON GEOSCIENCES, ENVIRONMENT, AND RESOURCES

M. GORDON WOLMAN, *Chair*, The Johns Hopkins University, Baltimore, Maryland
PATRICK R. ATKINS, Aluminum Company of America, Pittsburgh, Pennsylvania
EDITH BROWN WEISS, Georgetown University Law Center, Washington, D.C.
JAMES P. BRUCE, Canadian Climate Program Board, Ottawa, Canada
WILLIAM L. FISHER, University of Texas, Austin
EDWARD A. FRIEMAN, Scripps Institute of Oceanography, LaJolla, California
GEORGE M. HORNBERGER, University of Virginia, Charlottesville
W. BARCLAY KAMB, California Institute of Technology, Pasadena
PERRY L. MCCARTY, Stanford University, Stanford, California
S. GEORGE PHILANDER, Princeton University, New Jersey
RAYMOND A. PRICE, Queen's University at Kingston, Ontario
THOMAS C. SCHELLING, University of Maryland, College Park, Maryland
ELLEN K. SILBERGELD, Environmental Defense Fund, Washington, D.C.
STEVEN M. STANLEY, The Johns Hopkins University, Baltimore, Maryland
VICTORIA J. TSCHINKEL, Landers and Parsons, Tallahassee, Florida

Staff

STEPHEN RATTIEN, Executive Director
STEPHEN D. PARKER, Associate Executive Director
MORGAN GOPNIK, Assistant Executive Director
JEANETTE SPOON, Administrative Officer
SANDI FITZPATRICK, Administrative Associate

COMMISSION ON LIFE SCIENCES

THOMAS D. POLLARD, *Chair*, Johns Hopkins Medical School, Baltimore, Maryland
BRUCE N. AMES, University of California, Berkeley
JOHN C. BAILAR, III, McGill University, Montreal, Canada
J. MICHAEL BISHOP, University of California Medical Center, San Francisco
JOHN E. BURRIS, Marine Biological Laboratory, Woods Hole, Massachusetts
MICHAEL T. CLEGG, University of California, Riverside
GLENN A. CROSBY, Washington State University, Pullman
MARIAN E. KOSHLAND, University of California, Berkeley
RICHARD E. LENSKI, Michigan State University, East Lansing
EMIL A. PFITZER, Hoffmann-LaRoche, Inc., Nutley, New Jersey
MALCOLM C. PIKE, University of Southern California School of Medicine, Los Angeles
HENRY C. PITOT, III, University of Wisconsin, Madison
JONATHAN M. SAMET, University of New Mexico School of Medicine, Albuquerque
HAROLD M. SCHMECK, JR., Armonk, New York
CARLA J. SHATZ, University of California, Berkeley
SUSAN S. TAYLOR, University of California, San Diego, LaJolla
P. ROY VAGELOS, Merck & Company, Inc., Whitehouse Station, New Jersey
JOHN L. VANDEBERG, Southwest Foundation for Biomedical Research, San Antonio, Texas

Staff

PAUL GILMAN, Executive Director
SOLVEIG PADILLA, Administrative Assistant

The National Academy of Sciences is a private, nonprofit, self-perpetuating society of distinguished scholars engaged in scientific and engineering research, dedicated to the furtherance of science and technology and to their use for the general welfare. Upon the authority of the charter granted to it by the Congress in 1863, the Academy has a mandate that requires it to advise the federal government on scientific and technical matters. Dr. Bruce Alberts is president of the National Academy of Sciences.

The National Academy of Engineering was established in 1964, under the charter of the National Academy of Sciences, as a parallel organization of outstanding engineers. It is autonomous in its administration and in the selection of its members, sharing with the National Academy of Sciences the responsibility for advising the federal government. The National Academy of Engineering also sponsors engineering programs aimed at meeting national needs, encourages education and research, and recognizes the superior achievements of engineers. Dr. Robert M. White is president of the National Academy of Engineering.

The Institute of Medicine was established in 1970 by the National Academy of Sciences to secure the services of eminent members of appropriate professions in the examination of policy matters pertaining to the health of the public. The Institute acts under the responsibility given to the National Academy of Sciences by its congressional charter to be an adviser to the federal government and, upon its own initiative, to identify issues of medical care, research, and education. Dr. Kenneth I. Shine is president of the Institute of Medicine.

The National Research Council was organized by the National Academy of Sciences in 1916 to associate the broad community of science and technology with the Academy's purposes of furthering knowledge and advising the federal government. Functioning in accordance with general policies determined by the Academy, the Council has become the principal operating agency of both the National Academy of Sciences and the National Academy of Engineering in providing services to the government, the public, and the scientific and engineering communities. The Council is administered jointly by both Academies and the Institute of Medicine. Dr. Bruce Alberts and Dr. Robert M. White are chairman and vice chairman, respectively, of the National Research Council.

ACKNOWLEDGEMENTS

The membership of this committee has evolved since our review of EMAP began in 1990. We are grateful to the following former members for their contributions to the committee's work:

James Gosz, Chair (1990-1992)
Edwin H. Clark (1990-1993)
Arthur Cooper (1990-1992)
Shirley Dreiss (1990-1993)
Charles C. Johnson, Jr. (1990-1994)
Raymond A. Price (1990-1994)
Donald R. Strong (1990-1994)
John Sutherland (1990-1992)

In addition, we are very grateful to the following EPA officials and contractors for their many presentations and other assistance during our review of EMAP.

Darvene Adams
Craig Barber
H. Matthew Bills
Robert Currie
Tom DeMoss
Marge Holland
Robert Huggett
Hal Kibby
William Kepner
Rick Linthurst

Ed Martinko
Dan McKenzie
Jay Messer
Scott Overton
John Paul
Steve Paulsen
Paul Sandifer
Jerry Stober
Kevin Summers

Contents

EXECUTIVE SUMMARY 1

1 INTRODUCTION 11
    Background, 11
    The Present Study, 15

2 OVERALL ASSESSMENT 18
    Introduction, 18
    The EMAP Assessment Framework, 18
    Indicators, 24
    EMAP Sampling Density and Sampling Frequency, 27
    Statistics, 30
    Problems of Summarizing EMAP Data to Standard Federal Regions, 32
    Integration, 35
    Program Coordination Within EMAP, 39
    External Scientific Review, 41
    EMAP's Place in the Federal Government, 43

3 PROGRAM-WIDE COMPONENTS 46
    Landscape Characterization and Ecology, 46
    EMAP Indicator Development Strategy, 50
    Information System, 59

4 RESOURCE COMPONENTS 65
    Agroecosystems, 65
    Estuaries, 71
    Forests, 76
    Great Lakes, 79
    Surface Waters, 87

REFERENCES 99

APPENDIX A
    September 20, 1994 letter from Dr. Edward Martinko,
    Director, EMAP 105

APPENDIX B
    May 4, 1994 letter from Gary J. Foley, EPA Acting
    Assistant Administrator for Research and Development 119

APPENDIX C
    EMAP Documents Reviewed by NRC Committee 143

APPENDIX D
    Biographical Sketches of Committee Members 159

Executive Summary

EPA's Environmental Monitoring and Assessment Program (EMAP) was established to provide a comprehensive report card on the condition of the nation's ecological resources and to detect trends in the condition of those resources. At EPA's request, the National Research Council's Board on Environmental Studies and Toxicology and Water Science and Technology Board established the Committee to Review EPA's Environmental Monitoring and Assessment Program. This fourth and final report is the committee's overall evaluation of the program.

In 1988, the Science Advisory Board of the U.S. Environmental Protection Agency recommended that EPA "undertake research on techniques that can be used to help anticipate environmental problems," and that "an office be created within EPA for the purpose of evaluating environmental trends and assessing other predictors of potential environmental problems before they become acute".

Following the Science Advisory Board's advice, EPA established the Environmental Monitoring and Assessment Program (EMAP) "to monitor ecological status and trends, as well as to develop innovative methods to anticipate emerging environmental problems before they reach crisis proportions". In 1993 EMAP's stated goals were to:

1. Estimate the current status, trends, and changes in selected indicators of condition of the nation's ecological resources on a regional basis with known confidence.

2. Estimate the geographic coverage and extent of the nation's ecological resources with known confidence.

3. Seek associations between selected indicators of natural and human stresses and indicators of the condition of ecological resources.

4. Provide annual statistical summaries and periodic assessments of the nation's ecological resources.

As described by EPA, EMAP is unified by its approach to landscape characterization, the application of a coherent strategy for the choice and the development of indicators, and the use of a probability-based sampling approach that uses a hexagonal grid for identifying sampling sites. There are eight resource groups identified by the program: agroecosystems, arid (now rangeland) ecosystems, forests, the Great Lakes, estuaries, inland surface waters, wetlands (recently subsumed under surface waters and the Great Lakes), and landscape ecology. These resource groups are intended to represent ecosystem types or resources of national interest, and to provide a basis for incorporating ecological knowledge into the design of indicators and sampling programs.

The committee's reviews of other EMAP components such as forests and estuaries and surface waters were published as separate reports. The executive summaries of these reports are in Chapter 4.

After four years of review, the committee retains its belief that EMAP's goals are laudable. However, because achieving the goals of this ambitious program will require that EMAP successfully meet many difficult scientific, practical, and management challenges, the committee continues to question whether and how well all these goals can be achieved. This final report reiterates that general assessment.

Executive Summary

As first conceived and presented to the committee in 1991, EMAP was significantly different than it is today. Several of its central features and components seem to have less importance in mid-1994 than they did in 1991. The reverse is also true: the resource groups have become much more important and are leading the program. One of the major strengths of EMAP as initially presented was that it planned to integrate information across regions and across resource types, but the nature and extent of that integration is still not clear.

Given the need for 10 years or more of data to sample regions and distinguish trends, nobody—including the members of this committee—can be certain whether, or how fully, EMAP will achieve its stated goals. This is to be expected for a large, ambitious, and novel program like EMAP. However, the program-wide concerns expressed in the committee's previous reports, in Chapter 2 of this report, and summarized below, are so important that EMAP will have little chance of achieving its goals if they are not addressed. Concerns revolve around the following issues.

- The EMAP sampling program may operate at too coarse a scale in space and time to detect meaningful changes in the condition of ecological resources.
- EMAP's success will be diminished if it does not develop reliable, scientifically defensible indicators for measuring change. The development of indicators of ecological health or integrity and of aesthetic quality appear to be particularly challenging.
- EMAP's success will be diminished if it does not select the right assessment end points (i.e., the end effect that is the goal of the monitoring program), something it has not done so far.
- EMAP's success will be diminished if the retrospective or prospective monitoring approach does not match the assessment needs and the needs of policymakers.
- EMAP needs to incorporate the best scientific advice in the design, implementation, and review of its program.
- EMAP has not yet fulfilled its promise of innovation and national comprehensiveness because the programs to integrate

information across space, time and resource types have not been developed. The most important of these are an indicator-development strategy, information management, and landscape characterization.

• EMAP's information management system must support efficient access to a large, distributed database and application of an appropriate range of information processing tools.

• Lack of continuity in staffing at EMAP has inhibited development of the program. EMAP cannot succeed unless the government (i.e., the administration and the Congress) makes a sufficient financial commitment to EMAP to support administrative and technical excellence, continuity, and efficiency in program management. That commitment is necessary for EMAP to succeed, but is not sufficient by itself.

A September 1994 letter from EMAP director Edward Martinko (Appendix A) describes EMAP's recent responses to earlier NRC reports and provides additional updates about the program. Many of the changes described appear to be in line with the earlier committee recommendations. EMAP has not provided more detailed documentation of these encouraging changes, so this report has not been substantially altered. However, recommendations in this report that deal with matters directly addressed by Dr. Martinko's letter are indicated with an asterisk.

RECOMMENDATIONS

Statistics, Sampling, and Design

• **EMAP should consider design changes that would increase the probability of detecting smaller-scale ecological changes.** Some possibilities include increasing revisitation rates at a subset of sample sites; inclusion of a set of nonrandomly selected sentinel sites with intensive data-collection, such as the Long

Term Ecological Research (LTER)/Land Margin Ecosystems Research (LMER) networks; and stratified random sampling by ecoregion with data-quality objectives specified for strata. If EMAP does not adopt these design changes itself, then it should become extremely closely and explicitly coordinated with a program that has these features.

• **EMAP should consider further combining effects-oriented and stressor-oriented monitoring approaches.** Predictive, or stressor-oriented, monitoring seeks to detect the cause of an undesirable effect (a stressor) before the effect occurs or becomes serious. Retrospective, or effects-oriented, monitoring seeks to detect the effect after it has occurred. EMAP has relied mostly on the latter. Stressor-oriented monitoring will increase the probability of detecting meaningful ecological changes. As in the above point, if EMAP does not adopt these changes, it should become closely coordinated with a program that monitors in this way.

• **EMAP should undertake power analyses regarding the effectiveness of the sampling design for each resource group.*** A power analysis is an analysis of the statistical strength of an approach to detect change if a change exists. Different resource groups have adopted different sampling approaches. All the resource groups should adopt the practice of the EMAP lakes component, which has assigned teams of statisticians to assess the effectiveness of EMAP for that particular resource.

• **EMAP should reconsider its detection criterion of a 20 percent change over 10 years.** In some systems, such a large change is unlikely to occur in nature, while in other systems, a much smaller change would elicit concern. EMAP should also consider systems or indicators for which a change in the *variance*, rather than mean or median, is important.

---

*Recommendations marked with an asterisk are addressed in Dr. Martinko's 9/20/94 letter describing recent changes in EMAP.

Indicators

- **EMAP should initiate a major, focused research program on indicator development.*** Indicator development is at the heart of the EMAP program. Without a well-considered set of indicators for each resource group, EMAP will not fulfill its goal of presenting an evaluation of the nation's ecological resources. The difficulty and importance of indicator development requires EPA to attract the highest quality researchers in the environmental sciences to this program. The program should include a combination of internal research (by EMAP scientists) and external research involving open announcements of funding availability with peer-reviewed grants.

- **Each EMAP resource group should develop one or more *mechanistic* conceptual models of its resource, based on current scientific knowledge.*** These models should serve as explicit hypotheses about those aspects of ecosystem structure and functioning relevant to the assessment end points the group has chosen. The models must be detailed enough to include potential indicators, assessment end points, and key variables.

- **EMAP should provide program-wide guidance for numerous evaluation issues if the indicator-selection strategy is going to yield the nationally applicable set of indicators EMAP envisions.** The committee recommends as a high priority the explicit and early evaluation of the statistical properties of all potential indicators. Such evaluations should include analyses of the properties of the mean, variance, and behavior of the index in power tests. If this cannot be done analytically, then simulation analyses should be performed.

- **EMAP should carefully evaluate each potential indicator at incrementally larger spatial scales.** EMAP needs to make sure that it has information on the domains of usefulness of its indicators—at what spatial and temporal scales are they reliable, and at what scales are they less reliable? The ways in which the various resource groups deal with this problem will have important consequences for the selection of nationally implemented

Executive Summary 7

indicator metrics. Program-wide strategies for dealing with this issue should be developed now, in time to be applied with some uniformity across the resource groups.

Integration

- **EMAP should develop key integration and assessment questions that cross resources.** This would help focus the program and significantly extend its value nationwide.
- **EMAP should designate resources for integration.** As EMAP now stands, there are relatively few financial or other resources directed specifically at integration. Such resources could be directed in various ways, but several important needs must be met. Individual resource programs directed at integration must have access to the information management system, and must have computer and software resources to generate and test generalizations. One approach would support a team of individuals who focus on developing and addressing the integration and assessment questions, and who either work together at one physical location or were coordinated among resource groups by a central office. Key members of this group must be participants of the Landscape Characterization, Landscape Ecology, and Indicator Development groups. The new Integration and Assessment Program is a positive step in this direction, but we do not have a good description of the activities of this program.
- **EMAP should develop coordinated sampling between terrestrial, aquatic, and atmospheric resources.**[1] Resources

---

[1]*One committee member has been deeply concerned about the apparent lack of communication between senior administrators and possibly senior scientists, in the Air and Deposition Section of EMAP and those in major federal, state, and international agencies (e.g., Canada and Mexico) who are also heavily involved in ecological risk assessments and environmental protection. This (continued on p. 8)*

appearing to have very important ecological connections due to hydrologic linkages are now being sampled in separate locations. The design would be enhanced by a cooperative sampling scheme between resource groups in which lakes and streams were sampled in watersheds whose terrestrial systems (forests, agroecosystems, arid systems) also were being sampled. A stratified random system such as this would not compromise EMAP's ability to make regional-scale generalizations based on probability-based samples. The data sets would be considerably stronger because the spatial covariance of the data sets could be used to test hypotheses related to cause and effect relationships.

Possible examples include indicators reflecting net primary productivity, biological diversity, and aesthetic value. At present, it is unclear whether or not the assessment questions in each resource group are similar enough to lead to parallel sets of indicators. Such symmetry among resource groups, while not essential to basic EMAP objectives, would greatly enhance the scientific and analytical value of the data collected.

Appropriate Scale and Boundaries Of Regions

- **EMAP should choose ecologically meaningful units as the primary scale for summarizing and reporting data.** Ecologically meaningful units, such as Bailey's or Omernik's ecoregions, should be the primary objects of statistical analysis and data reporting rather than political units or EPA regions. In general,

---

member feels strongly that such inter- and indeed intra-agency interactions are essential for effective coordination of monitoring and assessment efforts involving the atmospheric transport, transformations, and deposition (as well as associated intermedia transport) of a wide range of hazardous gaseous and particulate air pollutants.

Executive Summary 9

EMAP should reconsider the scale and boundaries of units for which the national program summarizes and reports data.

Coordination And Management

- **EMAP is unlikely to succeed unless EPA commits permanent, senior-level positions to the program, and recruits qualified people to fill them.** Commitment and continuity are crucial for the implementation of such an innovative national program. Too many important responsibilities in EMAP have been assigned to people on temporary Interagency Personnel Agreements (IPAs) or to contractors.
- **The committee recommends that EPA senior administrators facilitate close working relationships between EMAP and appropriate offices and divisions of EPA, including other research programs in the Office of Research and Development.** In particular, EMAP should continue in its efforts to develop close working relationships with the EPA Office of Water to capture the benefits of EPA's past experience in collecting data on surface waters. Continued reliance on the experience of such programs leverages EMAP's resources and brings complementary expertise to the program.
- **EMAP should develop and maintain an administrative structure that demands close communication and interaction among EMAP-LC (Landscape Characterization), EMAP-IM (Information Management), and each of the resource groups.** This structure could take several forms, such as locating lead personnel of each of these groups at a central office or some other mechanism that requires regular communication among these groups.
- **The committee recommends that EMAP continue its efforts to coordinate its activities with those of other agencies.** The Memorandum of Understanding, signed by National Biological Service director H. Ron Pulliam and EPA Office of Research and Development director Robert Huggett (MOU, September 30,

1994) is an excellent example of such coordination. The committee encourages further efforts with programs like the National Water Quality Assessment of the U.S. Geological Survey.

External Scientific Review

- **The current external review structure of EMAP should be modified so that its core is a permanent panel, with rotating membership, to provide continuity.** A permanent board of accomplished scientists may provide more expertise and consistency of viewpoint than EMAP has had access to heretofore. The panel should advise both at the level of resource groups, such as the forests or estuary resource level, and at the level of the entire EMAP program.

Information Management

- **While top-down planning for EMAP's information system is important, EMAP should base such planning on the viewpoint that the information system is a scientific database system, rather than an information system focused on the needs of management if the Information Management System is to function and facilitate integration among research groups as envisioned by EMAP.** In particular, the planning should focus on the design of an environment that is sensitive to user requirements and that provides excellent hardware, software, and support personnel.

1

Introduction

BACKGROUND

In 1988, the Science Advisory Board of the U.S. Environmental Protection Agency recommended that EPA "undertake research on techniques that can be used to help anticipate environmental problems," and that "an office be created within EPA for the purpose of evaluating environmental trends and assessing other predictors of potential environmental problems before they become acute" (EPA Science Advisory Board, 1988).

Environmental regulations and management have been estimated to cost more than $100 billion per year in the United States (see NRC 1993a for estimates). Many environmental problems have diminished as a result of such expenditures—e.g., environmental lead, air and water pollution in many areas—but some have not and new potential and actual problems continue to arise. Clearly, there is a need for an assessment of the degree to which regulations and management in relation to natural resources have been worthwhile. The public needs to know the degree to which land-use programs are protecting our resources and if pollution-control measures are making a difference. It would also be helpful to know which programs are working best and which are less successful.

Following the Science Advisory Board's advice, EPA established the Environmental Monitoring and Assessment Program (EMAP) "to monitor ecological status and trends, as well as to develop innovative methods to anticipate emerging environmental problems before they reach crisis proportions" (EPA, 1991). In 1993 (EPA, 1993) EMAP's stated goals were to:

1. Estimate the current status, trends, and changes in selected indicators of condition of the nation's ecological resources on a regional basis with known confidence.
2. Estimate the geographic coverage and extent of the nation's ecological resources with known confidence.
3. Seek associations between selected indicators of natural and human stresses and indicators of the condition of ecological resources.
4. Provide annual statistical summaries and periodic assessments of the nation's ecological resources.

As described by EPA, EMAP is unified by its approach to landscape characterization, the application of a coherent strategy for the choice and the development of indicators, and the use of a probability-based sampling approach that uses a hexagonal grid for identifying sampling sites. There are eight resource groups identified by the program: agroecosystems, arid (now rangeland) ecosystems, forests, the Great Lakes, estuaries, inland surface waters, wetlands (recently subsumed under surface waters and the Great Lakes), and landscape ecology. These resource groups are intended to represent ecosystem types or resources of national interest, and to provide a basis for incorporating ecological knowledge into the design of indicators and sampling programs.

EMAP — Vision and Realities

The goal of EMAP from its inception has been to monitor and assess the condition of the nation's ecological resources to contribute to decisions on environmental protection and management. In the beginning, this mission was to be attained by determining the location and extent of the nation's ecological resources; establishing, with known confidence, status and trends in the condition of these resources; and assessing the relationships between the status and trends in condition and known stressors. Over the past five years, the methods for achieving EMAP's goals have developed into the four distinct objectives, described on the previous page.

EMAP will take an effects-oriented approach to monitoring, and will operate on a regional scale. According to EMAP documents, it will be capable of addressing such questions as: the proportion of the nation's lakes that are eutrophic, the changing area of geographic coverage of forest in the U.S., the proportion of fragmented landscapes in the southeast, and the proportion of the harbors and bays of the Great Lakes that are toxic to aquatic organisms.

EMAP will not address the following types of issues. EMAP is not a state-level program, so it will not explain the proportion of eutrophic lakes in any particular state. EMAP was designed to report on populations of resources rather than individual entities, so it will not answer questions about the condition of a particular lake. Since EMAP is not a cause and effect program (Thornton et al., 1994), it will not examine the causes of change (e.g. the relationship of agricultural practices to lake conditions). However, EMAP is able to associate regional changes in resources and stressors. For example, an association between the eutrophic status of mid-western lakes and the ban on phosphate in laundry detergents would be possible.

Various groups ranging from the EPA and Congress to the scientific community have held a wide variety of expectations of EMAP. It is helpful to review some of these expectations in light of what EMAP will probably be able to accomplish. EMAP as currently envisioned will be a broad-scale monitoring and assessment program. EMAP has stimulated scientific inquiry and will continue to do so; however, it is not itself a research program, and it probably will not directly add a great deal to our scientific understanding of ecological processes. Research should, however, play a role in EMAP in several direct ways. Research is essential for the development of indicators based on sound conceptual models; for the identification of specific indicators to assess; and in the screening and validation of indicators after they are selected. Sound scientific procedures will also be important in the assessment of data collected by EMAP, in the interpretation of the data, in linking them to policy decisions, and in identifying the need for the development of new indicators. As discussed elsewhere, the scientific community will play an essential role in the meaningful review of EMAP.

EMAP will provide insight into environmental policy questions, but will not provide answers. It will detect some environmental problems and suggest hypotheses as to their causes. EMAP is designed to collect and report information with a high level of known confidence for EPA's standard federal regions, but it will not provide information with the same level of precision for states, individual congressional districts, or ecological regions, although the EMAP design can be modified to provide such information. EMAP proposes to provide data on ecological indicators that have been chosen to provide "quantitative estimates of the condition of ecological resources, the magnitude of stress, the exposure of biological components to stress, or the amount of change in condition" (Barber, 1994).

EMAP will provide information on the extent and condition of the nation's ecological resources and will monitor trends in extent and condition. Not being a cause and effect program, EMAP will simply report changes in the environment rather than explain them. EPA can attempt to discover the causes of any adverse or beneficial changes EMAP reports. As currently conceived, the basic regional data collected by EMAP are only the first step in the complex process of providing the necessary information for making informed decisions on environmental protection and management. Additional information will be needed for informed policy decisions if and when status and trend data indicate a potential problem (Thornton et al., 1994).

Introduction

THE PRESENT STUDY

Committee Charge

In 1990, the National Research Council (NRC) appointed a committee to review EPA's Environmental Monitoring and Assessment Program (EMAP) at the request of the U.S. Environmental Protection Agency. The committee, which first met in March 1991, was charged with reviewing EMAP's overall design and objectives and considering ways to increase its effectiveness.

The NRC committee has reviewed approximately 150 EMAP documents (see Appendix B), and has been briefed by many EMAP officials including field personnel and the technical directors of EMAP's resource components.

The committee issued three reports prior to this final review of the program. Its first report, issued in 1992, supported the purpose and goals of EMAP, but raised substantive questions about the design and implementation. The committee issued this early review of the program in the hopes that the questions raised would be used to improve EMAP as it evolved. The committee also believed that the questions in the 1992 report could have been used by EPA as criteria for evaluating the results of pilot projects.

A second committee report issued in early 1994 reviewed the activities and plans of two EMAP resource groups: estuaries and forests (NRC, 1994a). This report concluded that much of the work of the estuaries and forests monitoring groups was well conceived and executed, and that many of the results of the demonstration projects were of considerable interest. The report also pointed out that the importance and uniqueness of EMAP depends on its being an integrated, coordinated, national program. The possibility of integrated descriptions of environmental trends across several resource types is what originally set EMAP apart from other intensive surveys in other agencies. However, this second report stated that no pilot studies had attempted any such integration by 1994, and that little thought had been given to the scientific underpinnings of cross-resource analysis.

The committee's third report, also issued in 1994 (NRC, 1994b), reviewed another EMAP resource group: EMAP-Surface Waters (EMAP-SW). In particular, the report reviewed and commented on the Lakes Pilot Project and on early information available on the streams program. The report commended EMAP-SW for its investigation into the critical ways different sources of variation will affect EMAP's ability to detect status and trends. The report states that EMAP-SW succeeded in organizing its implementation pilot and planning the logistics of the operation. The field portion of the regional assessment of the pilot was successful, and EMAP-SW gained valuable experience in the site selection process and in evaluating the logistical aspects of the program.

In general, however, the report on EMAP-SW stated that the pilot study needed substantial improvement. It failed to address all of its questions and goals and those goals and questions are a very incomplete list of the fundamental issues that need to be addressed before the surface waters program is ready for full implementation. In particular, issues of coordination among resource groups, relationships between indicators and specific stressors, and ability to make inferences on scales ranging from single lakes to entire regions, are not addressed by EMAP-SW. The committee also reviewed a stream pilot study and concluded that it was premature for EMAP to embark on the stream pilot study at this time because the sampling strategy is inadequate to characterize stream quality either chemically or biologically. This report also addressed the lack of oversight and involvement of senior scientists from a central management team at EMAP Center, which might have enhanced the scientific rigor of the pilot study.

Introduction 17

This Report: the Final Review

The present report represents the committee's judgments after four years of review. Chapter 2 focuses on matters that apply to the whole program, including the purposes of monitoring and those kinds of problems for which it is a more or less useful tool. Chapter 3 discusses the various components of EMAP that apply to all resource groups in EMAP, while Chapter 4 treats the resource components. In some cases, the committee has recently completed and published a review of the component, and only the executive summary of that report is reproduced, with any updates if they are applicable. The committee's conclusions and recommendations follow each chapter and the main ones are presented in the executive summary.

2

Overall Assessment

INTRODUCTION

EMAP is a program of many parts, but all the parts share common goals. The purpose of this chapter is to review those matters that apply to all parts of the program. They concern the design and implementation of monitoring; the sampling protocols; the development of indicators; integration among various parts of the program; program coordination within EMAP, within EPA, and within the federal government; and external scientific review.

THE EMAP ASSESSMENT FRAMEWORK

Most ecological monitoring programs are driven by some explicit or implicit set of assessment questions. EMAP is no different. The topic of this section is the basic monitoring approach taken by EMAP—retrospective monitoring. This well-established monitoring approach is suited for many environmental problems, but not for all; it is essential that any evaluation of EMAP have a clear view of those kinds of environmental problems that EMAP is likely to help identify and those for which other monitoring approaches would be better.

Retrospective or effects-oriented monitoring is monitoring that seeks to find effects by detecting changes in status or condition

Overall Assessment

of some organism, population, or community. Examples include monitoring the body temperature of a person, monitoring the productivity of a lake, monitoring the condition of foliage in forests, and so on. It is retrospective in that it is based on detecting an effect after it has occurred. It does not assume any knowledge of cause-effect relationships, although the intention is usually to try to establish a cause if an effect is found. It is EMAP's general approach.

Predictive or stress-oriented monitoring is monitoring that seeks to detect the known or suspected cause of an undesirable effect (a stressor) before the effect has had a chance to occur or to become serious. Examples include monitoring the cholesterol level in a person's blood, monitoring the stress level along a geological fault, monitoring animal tissues for the presence of known carcinogens or other disease-causing compounds, and monitoring with a canary the toxic gas level in a mine. It is predictive in that the cause-effect relationship is known, so that if the cause can be detected early, the effect can be predicted before it occurs.

The EMAP Assessment Framework (Thornton et al., 1994) is the formal exposition of the assessment context for the EMAP program. As such, it is the fundamental statement of the philosophical and practical requirements of EMAP's data-gathering activities. This is an especially critical element in the EMAP program development because in recent years EPA has been actively promoting ecological risk assessment (NRC, 1993b, RAF, 1992) as a new operating paradigm. Early EMAP documents were curiously silent about this larger EPA perspective, raising some question about the extent to which these initiatives were coordinated. In February 1994, EPA-EMAP released a document entitled *Environmental Monitoring and Assessment Program Assessment Framework* (Thornton et al., 1994). It is a welcome—if long overdue—addition to the voluminous descriptive literature EMAP has generated. It contains the most developed and lucid descriptions to date of the overall philosophy and approach behind the EMAP program. Together with the Indicator Development Strategy, the Assessment Framework clearly outlines the rationales, approach, and objectives of the EMAP program.

The Assessment Framework document describes the proposed role of EMAP in terms of EPA's ecological risk-assessment paradigm. It makes clear that the program is conceived of as having two distinctive roles in the ecological risk assessment process. As a supplement to the problem formulation phase of risk assessment, EMAP is to identify emerging problems that will require the attention of other programs within EPA to determine comparative ecological risk. EMAP's second task is to provide documentation of the success or failure of national risk-management decisions. By reporting on long-term trends in environmental status, EMAP is expected to provide data on the effectiveness of regulatory decisions and risk management by the agency. Its purpose in this regard is to provide data to evaluate policy, thus helping to close the loop in the iterative Ecological Risk Framework (RAF, 1992). The authors argue effectively that, in this sense, EMAP is really an integral part of EPA's new operating paradigm.

The Assessment Framework document is the appropriate place for a full exploration of the benefits and shortcomings of retrospective risk assessment. Unfortunately, it provides no wide-ranging examination of this issue. Given the proposed scale of EMAP, EPA has the responsibility to provide the public with a more detailed analysis.

One of the most important features of the Assessment Framework is the discussion of the implications of EPA's decision to use a retrospective (effects-oriented) instead of predictive (stress-oriented) assessment model. The Assessment Framework describes both approaches in some detail and refers to them as being complementary, arguing that retrospective analysis will **"become increasingly important as assessments of larger scale problems are conducted because it will become increasingly difficult to establish specific cause-effect relationships..."** Because the whole of the EMAP monitoring strategy is based on retrospective analysis, EMAP needs to present a more rigorous exposition of its rationale for this strategic choice. It is not clear that the quoted statement above is in fact true. For example, in the 1950s, the poor hatching success of birds of prey such as the bald eagle was a large-scale problem clearly caused by accumulation of DDT in the eggshells. It is significant that this cause-effect relationship between DDT and eggshell thickness was not established by a monitoring program such as EMAP, but by a research program driven by clearly stated hypotheses. Sound policy (the banning of DDT and subsequent recovery of eagles) was based on this cause-effect research.

Predictive and retrospective approaches are complementary, but they are not equally useful in every risk assessment. For situations in which the risk or consequence is severe (e.g., a pedestrian crossing a busy highway), effects-based risk assessment is an inappropriate strategy. When the risk is lower, either because the effect is weak, or it can be mitigated, retrospective analysis is appropriate and probably more cost-efficient (e.g., exposure badges in low radiation environments). The EMAP approach will be more useful for some types of ecological risks than for others. In considering the value of EMAP as a national monitoring program, it is important to understand what kinds of ecological risk EMAP is likely to provide useful data on, and what kinds of risks will require more traditional predictive, or stressor-based, analyses.

The Assessment Framework's assertion that spatial and temporal scales are the major variables determining whether or not EMAP can provide useful assessment data may not be correct. The usefulness of retrospective data in risk assessment at any scale varies with (1) the severity of risk, (2) the probability of detecting the effect (related to statistical power of sampling procedure), and (3) the time lag required for a mitigating response:

Usefulness = P(detection)/(severity x response lag time).

Application of this simple model to a list of real or potential environmental risks (Table 2-1) suggests that there are many risks for which retrospective analysis is appropriate, two for which it may

Table 2-1 Applicability of retrospective assessment to various environmental threats

Environmental Threat	Applicability of Retrospective Assessment	Reason
landscape alterations	appropriate	low severity, good detection
increased regional loadings of nitrogen, phosphorus, sulphur	appropriate	low severity, good detection
regional habitat declines	appropriate	low severity, good detection
chronic toxic contamination	appropriate	low severity, good detection
efficacy of cumulative policy	appropriate	low severity, good detection
point-source pollution	inappropriate?	poor detection
acid precipitation	inappropriate?	large mitigation lag
acute toxic contamination	inappropriate	high severity
exotic species effects	inappropriate	large mitigation lag
global warming	inappropriate	high severity, large mitigation lag
regional nuclear contamination	inappropriate	high severity
ozone depletion	inappropriate	high severity, large mitigation lag
biological extinctions	inappropriate	poor detection, infinite mitigation lag

Overall Assessment 23

be appropriate, and many for which it is inappropriate. (These examples are ways to think about appropriate monitoring strategies, individual areas might need more detailed analyses.) It is noteworthy that many high-profile environmental risks fall into the latter category. The EMAP approach is no panacea, and it is important that claims by the program and expectations of the public be realistic. In general, when the probability of detecting an effect is high and the cost of failing to detect is not extremely large, EMAP's effects-based monitoring can provide useful input to the Ecological Risk Assessment process as indicated in the Assessment Framework document (Thornton et al., 1994). However, when the cost of failing to detect an effect early is high traditional predictive risk assessment, which emphasizes stressor monitoring and modeling, is a clearly preferable strategy.

Implications for EMAP Design

The practical usefulness of EMAP's retrospective monitoring design depends on achieving a sufficiently high probability of detecting ecologically important effects. As discussed elsewhere in this and in previous reports of this committee, EMAP's current design might not have sufficient statistical power to detect *ecologically important* changes. The program's data-quality objective of detecting a change of 2 percent per year over 10 years in the mean of an indicator across a standard federal region can be achieved (ASA Committee on EMAP, 1992). As discussed later in this chapter, the pertinent scales for ecological processes are considerably finer, however, and the usefulness of EMAP will depend on its ability to detect these smaller scale changes. EMAP should consider design augmentations to increase the probability of detecting smaller scale ecological signals. Some possibilities include:

- Increased revisitation rates at a subset of sample sites.
- Inclusion of a set of nonrandomly selected sentinel sites with intensive data-collection, such as the Long-Term Ecological Research network.

• Stratified random sampling by ecoregion with data-quality objectives specified for strata.

Hybrid Assessment Models

The Assessment Framework addresses the issue of enhancing the probability of detecting effects by calling for selection of indicators that are linked to specific environmental values. This has been termed "stressor-cognitive" indicator selection by EMAP staff. In a sense the EMAP indicator-development strategy leans towards including some aspects of a prospective assessment, or stressor-oriented, approach as well. In some of the resource groups, e.g., estuaries, this approach has been strongly implemented, with some of the indicators chosen to detect the most probable and common stresses (organic loading and toxic contamination). EMAP should consider further developing this hybridization of effects-oriented and stressor-oriented monitoring approaches. By focusing on indicators that are sensitive to the most likely known stressors for each resource group, an increase in the probability of detecting meaningful ecological changes may be achievable within any given sampling program.

There remain issues of the EMAP design and assessment approach that have not been adequately addressed such as: the ability to detect changes at an appropriate scale, whether the sampling return period is adequate to detect cyclical events, and the efficacy of specific "stressor cognitive" indicators. EMAP should continue to evaluate features of the design in the light of its ability to detect important ecological change in a meaningful, timely, and useful manner.

INDICATORS

A fundamental premise of EMAP is that the status of large and complex ecological systems can be assessed and monitored using a limited set of indicators. Choosing appropriate indicators has been a major focus of EMAP activity since the program began. Despite the obvious centrality of indicator development to EMAP, the completion of a comprehensive indicator-strategy document has been slow in coming. An early version of a strategy document (Olsen, 1992) was withdrawn following a major program reorganization. A new indicator-strategy document was developed more recently and was received by the committee in the spring of 1994 (Barber, 1994).

A strong reliance on biological measurements is highly appropriate for a monitoring program with EMAP's goals. Because environmental managers increasingly emphasize issues of biological integrity and integrated approaches to watershed and ecosystem management, environmental monitoring programs can no longer rely solely or primarily on measuring physical and chemical conditions of ecosystem quality.

In contrast to chemical indicators, which tend to reflect short-term or instantaneous conditions, biological indicators integrate conditions over time. This attribute is especially important for a monitoring program in which resource units are sampled at most once per year. In addition, advances in the development of multi-species biological indices using multivariate statistical methods offer some hope of defining and quantifying certain aspects of ecosystem status, and perhaps will allow useful quantification of the still nebulous concept of appropriate biological diversity, and the even more indeterminate concepts of ecological health and biological integrity.

Exciting advances offer opportunities for development of innovative, molecular-level measurements of ecosystem functioning and responses to stress. Examples include the use of genetic markers (gene probes) to detect the presence of certain organisms or types of organisms, molecular indicators to detect the exposure of organisms to classes of toxic compounds, and molecular measures of biological functions. Moreover, EPA needs information about biota in the nation's ecosystems that goes beyond questions of biological diversity and structural aspects of biological integrity. For example, there is a long-term need to test biota for levels of toxic contaminants, not only because of human health issues but also because of concerns about the health of animals that accumulate contaminants from lower levels of the food web (e.g., DDT and birds of prey).

On the other hand, there are several serious difficulties that must be overcome before EMAP is able to place more reliance on biological indicators in its long-term monitoring and assessment strategy. Many of these problems have been recognized by EMAP scientists in various documents discussing selection of indicators for specific resource groups (see Appendix A) and in the recent indicator-strategy document. The following are among the most important of these difficulties. First, virtually every characteristic of biological integrity one can think of has an extremely wide range of "appropriate values," depending on the nature of the ecosystem being considered. There are few, if any, absolute biological characteristics that identify an ecosystem as healthy, pristine, or undegraded. Beginning in 1991, the EMAP-Estuaries pilot study used an index that is compared to a local reference system (NRC, 1994a). This approach could be extended to other parts of EMAP. The extent to which a system is degraded is decided by comparison with similar systems. Second, biological variables do not respond in simple or linear ways to stress, creating special statistical difficulties in developing indicators for them. Third, many community and ecosystem-level measures of ecosystem function are quite insensitive to stress. Fourth, from a practical perspective, taxonomic identifications can be tedious, time-consuming, and expensive, thus perhaps limiting the number of samples that can be measured.

Finally, the most crucial problem is that for most ecosystems there are no quantifiable biological indicators of ecosystem health, biological integrity, or of several other societal values associated with ecosystems. A major research and development effort will be needed before workable, reliable, and cost-effective measures are available. To date, EMAP has not developed such an aggressive and comprehensive research program for development of biological indicators, and this is perhaps the most important research need facing EMAP.

Given the scope of the problem, EMAP should initiate a major, focused research program on indicator development. This is a formidable scientific challenge. If it can be successfully met, it will need the involvement of many scientists for a period of at least 5 to 10 years. It should include a combination of research by EMAP scientists and external research that involves open

Overall Assessment

announcements of funding availability with peer-reviewed grants. The difficulty and importance of this research requires that EPA attract the highest quality researchers in the environmental sciences to this program. Some of the needed research and development can be accomplished in association with existing pilot monitoring activities, but fundamental research also will be needed that cannot be addressed by sampling at the spatial and temporal scales of the pilot programs. As part of this indicator development program, scientists will need to study the relationship of the indicators being considered to assessment end points and the statistical properties and power of potential indicators. The research program should be directed not only at development of biological indicators appropriate for the various resource groups but also at applying new advances to develop new measures of ecosystem function as well as ecosystem structure.

EMAP SAMPLING DENSITY AND SAMPLING FREQUENCY

EMAP has developed a probability-based sampling design to address certain questions regarding the status and trends of ecological resource populations of interest in the United States. EMAP focuses on regional surveys rather than a "sentinel-site" approach; an EMAP objective is to select samples that are spatially well distributed so that the results will apply to fairly large areas. EMAP is willing to forego intensive, site-specific monitoring data in return for reliable data on regional changes and trends in regional population statistics (Messer et al., 1991). According to Foley (1994), assessment of temporal trends using sentinel-site-based monitoring will be left to "other ORD [EPA's Office of Research and Development], federal, and academic programs, because according to EPA, site-based information alone is insufficient for detecting 'meaningful trends at scales relevant to policy decisions'."

The EMAP hexagonal, grid design does make it possible to sample at varying spatial densities such that the base grid is a subset of higher density grids (Stevens, 1993). Sites will be revisited every four years. EMAP claims that the 4-year interval is consistent with the time scale of trends that EMAP must detect. "Trends that result in immediate extreme changes will be reflected in annual estimates, by other monitoring programs, or by casual observation. However, faint trends require some time before the cumulative change is detectable, and as great a population coverage as possible is needed in order to isolate subpopulations that may respond differently than others" (Stevens, 1993, p. 20).

A "large-scale, regional estimates only" approach is not compatible with one that claims to use the same spatial grid density to isolate sensitive subpopulations. By enhancing the grid density, such subpopulations might be identified and monitored. But they would have to be identified in advance by some other program or programs outside the EMAP sampling frameworks, because monitoring an enhanced grid at a regional or national scale would be prohibitively expensive. For instance, the selection of indicators of biological condition for the various resource groups may be driven more by what can be detected in four years' time than by their ecological relevance. Regarding the issues of detection of temporal or spatial processes outside the current EMAP sampling frameworks, EMAP claims that other federal programs (e.g., LTER, LMER) and state and regional programs including the Regional Environmental Monitoring Assessment Program (REMAP) are really the programs designed to investigate those issues. In that case, the coordination between EMAP and other more site-specific programs should be greatly enhanced.

Initially, EMAP had planned to augment the interpenetrating, 4-year-cycle design by sampling some sites annually; the contribution of this added component is described by Urquhart et al. (1993). Annual sampling during the initial years of the survey was mentioned specifically for lakes and streams as a way to increase the power of the EMAP design to detect trends (Larsen et al., 1993). As a result of further simulation studies, this augmentation sampling plan apparently was dropped and no lakes will be revisited annually for more than two consecutive years. This has the consequence of generating no site-specific information for longer than two years; the EMAP claim is that more information is to be gained by sampling more sites than by extensive repeat sampling at a single site (Appendix A).

Overall Assessment

The amount of statistical work assessing the effectiveness of the overall EMAP sampling design varies greatly among the different resource groups. For example, much work has gone into assessing the effectiveness of the EMAP-Surface Waters sampling design for lakes. The inclusion probabilities for the sampling design have been adjusted to yield an adequate sample of large lakes. Power analyses, sample size calculations, least significant trends, and investigation of variance components have been done for certain chemical and physical responses (Urquhart et al., 1993; Stehman and Overton, 1994). The EMAP lakes component has benefited from having teams of statisticians working specifically on assessing the effectiveness of EMAP for that particular resource. Similar work should be conducted in the other resource groups.

EMAP-Estuaries implemented a stratified sampling design in pilot demonstration projects to ensure adequate representation of small estuaries and tidal rivers, large tidal rivers, and large estuaries. EMAP for streams will not use the grid sample, but a probability sample where streams are treated as a discrete resource. EMAP-Surface Waters focuses on the population of stream miles rather than stream reaches; thus streams are sampled in proportion to their lengths. EMAP needs to make clear how such a sampling scheme will be compatible with a watershed approach to sampling. For example, it is not evident how stream sampling will be coordinated with the sampling of surrounding forest, or if the Forest Health Monitoring program is following the EMAP grid.

With respect to agroecosystems, the North Carolina pilot study compared the sampling design of the National Agricultural Statistical Survey (NASS, a well-established program) to the EMAP sampling design. The conclusion was that the EMAP sampling design is no more cost-effective than the NASS sampling design. This experiment is being repeated in the Nebraska agroecosystem pilot, but no reports have been released. Ultimately, the EMAP agroecosystem sampling design is expected to coordinate with the NASS design by augmenting the sampling grid and data requirements of NASS (further details in Chapter 4).

The statistical power to detect spatial or temporal differences (or even to answer specific questions regarding status) raises concerns because of economic and other practical constraints on the number of samples imposed on EMAP. For example, to obtain the acidification status of lakes in the Northeast, the original EMAP grid had to be enhanced beyond the base design. The sampling approach varies among the different resource groups; for this reason power analyses should be applied to each resource group to analyze the effectiveness of the sample design.

STATISTICS

EMAP will use cumulative distribution functions to display information in the annual statistical summaries. This approach does not focus on a single parameter, like a mean or median, but instead allows a compact display of an entire probability distribution for an indicator of interest. Changes in inherent variability, or in the tails of the distribution (e.g., the 90th or 10th percentiles), can be diagnosed using cumulative distribution functions, which will also have confidence bands around them. With increased numbers of samples, the level of confidence will not change, but the width of the confidence bands will become smaller. The power analyses that the committee has seen thus far deal with changes in the mean, and future power analysis research assessing the effectiveness of EMAP designs should pay attention to more extreme quantiles, like the 10th or 90th percentiles. This is especially important in light of the statement that confidence bounds obtained using the Horvitz-Thompson variance estimators (part of the statistical underpinnings of the EMAP grid design) are based on normal approximations. These normal approximations may be inadequate, even for moderate sample sizes, for estimating confidence bounds at the tails of the distribution (Lesser and Overton, 1994). For environmental monitoring objectives, the tails of the distribution may be where much of the interest lies for certain indicators.

EMAP now seems to have an extensive cooperative statistical research program, involving statistics departments in at least eight universities, and with at least twelve principal investigators. It is to EMAP's credit that many papers regarding the statistical underpinnings of this complex effort are now available in peer-reviewed literature.

Overall Assessment

Statistical analyses have varied according to the needs of the various resource groups. Nonetheless, every sampling site will generate multivariate data when sampled for various indicators. EMAP has considered the development of indices as a way of condensing multivariate data into single numbers for easier understanding and easier analyses. Rather than relying exclusively upon an index with unknown statistical properties however, it is possible to use the vector of original responses and apply multivariate statistical techniques for analysis, or exploratory techniques involving visualization of multi-dimensional data (e.g., Becker et al., 1987; Cleveland and McGill, 1988). Nonparametric multivariate techniques to ascertain differences among groups of data points and to detect trends in multivariate data also exist (e.g., Zimmerman et al., 1985; Saila, 1993). In response to a recommendation from this committee that nonlinear types of trends be investigated (including threshold effects and step functions) (NRC, 1994a), nonlinear trends are being included in the statistical research (Foley, 1994). There is value in using a multi-pronged data analysis approach, as one can have more faith when similar conclusions are derived from independent statistical methods.

Trends

In analyzing its ability to detect temporal trends, EMAP has used as its criterion for success the ability to detect a monotonic trend of 20 percent over 10 years. EMAP should consider in detail and in depth the conditions under which this criterion is likely to be useful, and those conditions under which it might not be meaningful. Are there some systems or some indicators in which a change in value of 20 percent over 10 years is so rare that it will likely never occur, or in which one might be extremely concerned about a much smaller change? Included in this analysis should be consideration of systems or indicators for which a change in the *variance* rather than location parameters (e.g., mean, median) is important. Measures of central tendency are insensitive to some types of changes in living systems. In studies of human and animal behavior for example, changes in distribution of reactions in the face of stress can be striking even while the mean or median remain more or less unchanged. While information on variability is contained in cumulative distribution functions, the power analyses thus far have concentrated on trends in the mean.

For example, a monitoring program that could detect only a 20 percent or greater change in mean annual surface temperatures in the United States, or in tropospheric ozone, or in human age-specific mortality rates, would not be sensitive enough to be useful to policy makers. Similarly, a 20 percent change in the visibility of lake waters over 10 years on a regional basis is much larger than one might reasonably expect to occur, and policies will need to be made before such a change can be detected. On the other hand, a 20 percent change in dissolved oxygen in estuaries or harbors appears to be small enough to be possible and important enough to be worth detecting.

PROBLEMS OF SUMMARIZING EMAP DATA TO STANDARD FEDERAL REGIONS

EMAP will collect data on a large number of grid points. The status and trends of condition of ecological resources can be reported on these or any other level. Some aggregation of data is required to yield the means and confidence intervals desired. EMAP intends to use Standard Federal Regions (SFRs) as the primary scale for summarizing data and inferring the status and trends of the nation's ecological resources.

Standard Federal Regions have a few advantages for data analysis and reporting. The major advantage is that EPA is administratively organized according to these regions. Summarizing EMAP data accordingly will facilitate its use by EPA field personnel as well as administrative personnel in Washington. In some cases, summarizing data by SFRs may help reporting to congressional delegations, particularly when congressional delegations themselves are organized accordingly. For example, New England is in SFR 1 and there may be times when EMAP wishes to report to the New England congressional delegation on the status and

Overall Assessment

trends of ecological resources in that region. However, not all congressional delegations are organized by SFRs.

In contrast, there are a number of serious ecological disadvantages and additional administrative difficulties in summarizing data by SFRs. First, these regions do not correspond to scales of relevant ecological processes. Ecological processes tend to occur at scales typical of first- or second-order watersheds, insect outbreaks, fires, and the like. This means areas of tens to thousands of hectares (typically best represented by maps at scales of 1:1,000 to 1:20,000). Therefore, summarizing data at the scale of SFRs—hundreds of thousands of square kilometers—will often obscure problems that arise at the ecosystem level. Second, the data-quality objectives set by EMAP SFRs will not be achieved at the smaller scales relevant to important ecological changes. Third, the boundaries of SFRs are not congruent with the boundaries of ecological units such as communities, ecoregions, or biomes. Ecological boundaries are generally determined by biological responses to geomorphic or climatic boundaries, not political boundaries. Also, SFRs are often not ecologically homogeneous; some western regions, for example, each include prairie, desert, montane forests, and alpine tundra.

An additional difficulty is that other federal agencies, the U.S. Forest Service, Soil Conservation Service, U.S. Fish and Wildlife Service, National Biological Service, and the National Park Service, are not organized by Standard Federal Regions as is EPA. Therefore, summarizing data in this way will impair information transfer to other land-management agencies.

Some of the problems of reporting data on SFRs can be illustrated by the EMAP Forests data summaries. The 1992 Forest Health Monitoring Statistical Summary reaches the following conclusions:

- SFR 1 (CT, ME, MA, NH, RI, VT) and 2 (NJ, PR, VI) combined have more dead trees (on a per-area basis) than either SFR 3 (DE, DC, MD, PA, VA) or 4 (AL, FL, GA, KY, MS, NC, SC, TN).
- Species density was used as a measure of species diversity of trees and saplings in SFR 1 and 2 combined, SFR 3, and SFR 4. Two species per unit area were used as a preliminary suboptimal threshold.

- SFR 4 had a significantly higher proportion of plots with suboptimal tree species density than SFR 3. SFR 1 and 2 combined and SFR 3 did not differ significantly for these proportions.
- SFR 1 and 2 combined had a significantly higher proportion of plots with suboptimal sapling species density than SFR 3 or 4.

The regions being compared are made up of widely different biomes and ecosystems, and so the comparisons are not informative. The potential reasons for the first finding are legion: Perhaps trees decay faster in the South, SFR 4; Perhaps forests in SFR 1 and 2 are, on average, older than those in SFR 3 or 4; Perhaps there is a higher pollution level in SFRs 1 and 2 than in SFR 3 or 4. This finding will be reason for alarm to a few, an interesting conundrum to many, and an example of "federal fog" to others. Such generalized conclusions are not good for EMAP nor will they engender much support for environmental monitoring.

The second and third findings are also problematic. Southern pine forests, which predominate in SFR 4, are naturally comprised of few species, but it is doubtful that they are in suboptimal ecological condition when compared to more species-rich high-graded woodlots, which predominate in parts of SFRs 1, 2, and 3. Likewise, climax forests in portions of SFR 1 and 2, which are in potentially optimal ecological condition, often have understories dominated by a single species—sugar maple. Broad generalizations such as findings 2 and 3 above are misleading, and by presenting easy targets for criticism, they will make it more difficult for EMAP to reach its objectives. The fourth finding is also problematic because optimal sapling density depends on species composition of specific stands, the ratio of ingrowth from seedling to sapling classes, mortality over all size classes, and the management objectives to be realized over a given planning horizon. Optimal sapling density is meaningless at regional scales.

The committee recognizes that the EMAP sampling design is flexible enough that data can be summarized in many different ways. However, the program's data-quality objectives pertain only to the primary scale for summarizing and reporting data, the SFRs. It is important that EMAP choose ecologically meaningful units as the primary scale on which to base data-quality objectives and for summarizing and reporting data. Ecologically meaningful units, such as Bailey's or Omernik's ecoregions, should be the primary objects of statistical analysis and data reporting. Summarizing data at this level will facilitate the development of meaningful hypotheses regarding causes and effects. Summarizing data by ecoregions also will allow users to investigate the geographic pattern of responses to stresses in a way that lends itself to further investigations. For example, is a particular ecoregion responding to a regional or global stress such as acid deposition or global warming in the same way throughout its range? If not, why not? Such questions are meaningless if the data are primarily summarized by SFR. Ecological data that are focused on finer scales could be summarized for SFRs secondarily, and the data would then meet or exceed EMAP's data-quality objectives.

The Regional Environmental Monitoring and Assessment Program (REMAP) operates at scales and with boundaries more appropriate to ecological processes. REMAP was not a part of the original EMAP concept, but it has become a potentially significant contribution. It uses EMAP indicators and the EMAP probability-based sampling scheme at an intensified sampling density to address local problems. EMAP should reconsider the scale at which the national program collects and reports data.

INTEGRATION

Need for Integration

As the first national-scale, multiresource monitoring program, EMAP represents a potentially significant addition to the myriad environmental monitoring programs run by other agencies. Two aspects of EMAP make it unique: the probability-based sampling design, and the inclusion of all of the biomes in the coterminous United States. In initial EMAP literature, this second aspect, inclusiveness of all resources, was the major justification for a new environmental monitoring program. The committee believes that integration among EMAP resource groups is crucial for the following reasons.

- To a large extent, major trends within resource types may only be explained by interactions among resources. For example, most pollution problems occur across resource groups defined by EMAP. An explanation of a forest's condition may require the assessment of regional, urban or agricultural practices as they influence nitrogen gas emissions. Virtually every national environmental issue to date has involved such connections. Integration among resource groups explains trends in ecological status that are controlled by spatially explicit source-sink relationships.
- A large, national, cross-resource monitoring program could lead to important new advances in our knowledge of interactions among resources.

Demonstrations of Integration

EMAP has given several indications that it plans to carry out integration among resource programs in the future. The committee has been informed that the resource groups are developing plans to use common methods. The fact that EMAP has or is developing a central sampling design, indicator development strategy, and information management system also suggests that EMAP is planning for future integration. Additionally, the presence of EMAP Center as a central administration provides support for integration. Perhaps most promising, the new Integration and Assessment Program suggests that major new advances can be made with the integration of the program.

Concerns

The committee recognizes that a large, complex program like EMAP must continue to develop and evolve, and that this evolution is a process of iteration between the conceptual, integrative elements (top-down directives), and the empirical, on-the-ground, resource group elements (bottom-up guidance). The top-down directives of the program appear to have weakened considerably since this review began. In 1990, EMAP representatives stated that the Landscape Characterization program and the Indicator

Overall Assessment 37

Development Strategy were the foundations of the integration program for EMAP. Since that time, the Landscape Characterization program has been changed substantially, such that now it appears to be focused entirely on land-cover mapping. The original elements that provided integration appear to be gone. These elements included guidance to resource groups on appropriate sampling intensity and locations, and analysis of indicators across resources. Finally, the technical support for cross-resource integration—the information system—still has not been fully developed. In sum, most of the positive aspects of integration listed above are not yet being implemented.

Recommendations

Coordinated Sampling Between Terrestrial, Aquatic, and Atmospheric Resources[1]

The groups are now sampling in separate locations resources that would appear to have very important connections because of their hydrologic linkages. The design would be much enhanced by a cooperative sampling scheme among resource groups in which lakes and streams were sampled in watersheds whose terrestrial systems (forests, agroecosystems, or arid systems) also were being sampled. A stratified random system such as this

---

[1]*One committee member has been deeply concerned about the apparent lack of communication between senior administrators and possibly senior scientists, in the Air and Deposition Section of EMAP and those in major federal, state, and international agencies (e.g., Canada and Mexico) who are also heavily involved in ecological risk assessments and environmental protection. This member feels strongly that such inter- and indeed intra-agency interactions are essential for effective coordination of monitoring and assessment efforts involving the atmospheric transport, transformations, and deposition (as well as associated intermedia transport) of a wide range of hazardous gaseous and particulate air pollutants.*

would not compromise EMAP's ability to make regional-scale generalizations, or to base those generalizations on probability-based samples. The data sets would be considerably stronger because the spatial covariance of the data sets could be used to test hypotheses related to cause and effect relationships. Several decades of measurements may be required to test these relationships on temporal data. In addition, the development of integration and assessment questions that cross resource groups could focus the program and significantly extend its value nationwide.

Designate Resources for Integration

As EMAP now stands, there are relatively few financial and human resources directed specifically at integration. Such resources could be directed in various ways, but several important needs must be met. Programs directed at integration must have access to the information management system, and must have computer and software resources to generate and test generalizations.

One approach would support a team of individuals that focused on developing and addressing the integration and assessment questions. The team would either have one physical location or would be coordinated among resource groups by a central office. Key members of this group would include participants of the Landscape Characterization, Landscape Ecology, and Indicator Development groups. The new Integration and Assessment Program is a positive step in this direction, but there is no description to date of the activities of this program.

A second approach would be to provide the funding for Request for Proposals to be extended to the scientific community. This approach could be implemented in a number of ways, either by targeting specific assessment questions, or by allowing the scientific community to develop the more general questions of most interest to EMAP. An intriguing possibility would include EMAP support (both financial and data support) of the new National Science Foundation Center for Ecological Analysis and Synthesis. The center is intended to serve as a resource for the

Overall Assessment

ecological community at large, and accessibility to EMAP data and technology would likely provide a great incentive for cross-resource analysis, assessment of ecology-policy links, and analysis of long-term and large-scale ecological status and trends.

PROGRAM COORDINATION WITHIN EMAP

EMAP has improved its internal coordination over the past several years. The creation of EMAP Center has resulted in a concentration of technical personnel who provide support to the resource groups. During the reviews of the Surface Water, Forest, and Estuary programs, and during meetings with the NRC committee, EMAP staff commented favorably on coordination between EMAP Center and the resource groups. The recent publication of a newsletter facilitates communication among all components of EMAP as well as those outside of EMAP. Nonetheless, improvement is needed.

Several features of EMAP inhibit internal coordination. EMAP personnel are dispersed in laboratories and offices across the country. Personnel in each resource group are typically clustered together, but they are separate from other resource groups and from EMAP Center. In addition, the autonomy of resource groups appears nearly absolute. The combination of geographic separation and administrative autonomy makes internal coordination difficult at best. The committee recommends that EPA consider the advantages of assigning all EMAP administrative personnel to a common location.

Report review within EMAP appears to be cumbersome. Reports delivered to the committee for review are typically based on information gathered two or more years previously. During the early stages of a new program, it is especially important to learn from experience and to modify the program based on the lessons learned. It is not clear that experience is used in a timely and efficient manner to improve subsequent work. Possibly, the slowness of report reviews is at least partially responsible for such inefficiency. EPA should consider means by which report generation and review can be speeded up. Centralization of personnel, reports in peer-reviewed journals, use of external reviewers, reliance on scientific criteria for reviews, and reduced sensitivity to being wrong and to the prevailing political philosophy are potential means to shorten the time for report review.

The rate of turnover in personnel appears high in comparison with other scientific programs and activities. EMAP has had three directors, three associate directors are leaving, and several resource and support groups have had multiple leaders. Some of this turnover is inevitable as positions are filled by contract employees with statutory limits on service. For example, the director and the coordinator of the Agroecosystem Resource Group serve under term-limited Interagency Personnel Agreements. Such turnover further disrupts coordination among EPA personnel. The committee questions whether contract employees can be as effective as direct employees in facilitating coordination within EPA. Individual relationships are the basis of collegiality and collaboration. Relationships between EPA employees and term-limited contract employees are frequently disrupted, inhibiting coordination. All federal agencies have complex hiring requirements, but EPA must address these issues. EPA should commit the senior-level positions required to assure continuity within the management of this important program and recruit qualified people to fill them.

Coordination between EMAP and other parts of EPA

Through its REMAP program, EMAP has demonstrated excellent coordination with its regional offices. Representatives from regional offices report that they have found the EMAP design and staff useful in assessing issues within the regions. EMAP should be commended for its efforts in finding the means to interact with its regional offices for mutual benefit.

EPA senior management should facilitate close working relationships between EMAP and appropriate offices and divisions of EPA. For example, EMAP should continue its efforts to develop close working relationships with the Water Office of EPA to capture the benefits of EPA's past experience in collecting data on surface waters. In addition, EMAP should coordinate closely with other research programs in the Office of Research and Develop-

Overall Assessment 41

ment. EMAP's partnership with the Exploratory Research Program in administration of the solicitation for proposals for indicator development has been productive. Continued reliance on the experience of such programs leverages EMAP's resources and brings complementary expertise to the program.

EXTERNAL SCIENTIFIC REVIEW

EMAP is attempting to carry out a number of approaches that are new to ecology, including using indicators of ecological status or health, developing and applying indicators in many types of ecosystems, and making measurements over an entire continent. All of these novel approaches require the best possible judgment from the most experienced researchers. Is EMAP receiving the external scientific review it needs to carry out its mission?

All parts of EMAP have brought in external reviewers at various stages of planning and implementation. For example, the indicator development program held extensive workshops of EPA contractors and independent scientists to come up with ideas about possible indicators. Panels of experts have advised the various resource groups, and the EMAP reports have received extensive review by up to 100 reviewers, many of them from outside of EPA. One recent activity, an EMAP-initiated funding of projects to develop new and better ecological indicators, used a National Science Foundation panel for scientific review and ranking of the proposed projects.

One successful use of external reviewers was carried out at the early stages of EMAP-Estuaries. At the request of EMAP, the Estuarine Research Federation (ERF) set up an expert panel that was briefed by EMAP and produced reports based upon three meetings held a year apart. The ERF panel and EMAP scientists met at the end of each year to review progress and discuss problems. Participants report that the advice from the ERF panel was taken seriously. Several years later, the estuaries panel of this NRC committee found that the ERF reports dealt very well with the strengths and weaknesses of the EMAP program.

EMAP has been willing to use external reviewers more than any other agency program. Nonetheless, the review process could be improved. First, it is important to use external reviewers as *advisors* as well as reviewers. The advice of these external reviewers should be used in designing parts of the program and its overall structure, as well as in reviewing finished or partly finished activities.

Second, at least in some cases, the reviewers could have been chosen better. For example, however skilled and knowledgeable EMAP contractors and employees might be, inasmuch as they have helped to design or implement the program, they are not independent of it, and therefore they should not be used as reviewers. Also, there is value to having reviewers who are active in other, related fields and whose funding and scientific publications indicate that they are leaders in developing the scientific research and technologies that EMAP needs.

Third, EMAP seems to have convened a new review panel for each task. Given the complexity of the EMAP program, it is a major task to bring each panel up to the proper level of understanding. Permanent or long-term scientific review panels for each of the eight resource groups were a part of the EMAP plan, but have not yet been set up. In addition to the need for good external review of these various resource groups and funding initiatives, there is also a need for overview of the entire program that is not filled by reviews of each separate part.

The current external review structure of EMAP should be modified so that its core is a permanent panel, with rotating membership, to provide continuity. This panel would advise both at the level of resource groups, such as the arid land or estuary resource level, and at the level of the entire EMAP program. The advice will be taken more seriously if good relationships are established between members of the panel and the directors of the various sections of EMAP. Above all, best scientists have to be recruited as advisors. Advice about membership should be sought from scientific societies, as was done with the Estuarine Research Federation, and from the program managers at the National Science Foundation, the Department of Agriculture, and the Department of Energy who organize competitive research programs in ecology. A structure for permanent advisory panels was proposed by EMAP personnel but has not been implemented.

Overall Assessment 43

EMAP's PLACE IN THE FEDERAL GOVERNMENT

The Administrative Setting for EMAP

EMAP is a large program that will require an appropriate administrative setting to be successful. It is important to reflect on the extent to which EPA has provided an appropriate administrative environment for EMAP.

Characteristics of the ideal administrative setting that would enhance EMAP's chances of success are described below. Of course, no agency has all these characteristics. Indeed, some of the characteristics are somewhat incompatible with each other. Conversely, some of the characteristics are related to each other—the first two, for example.

- EMAP's parent agency should be nonregulatory. A monitoring program housed within a regulatory agency will face problems of internal conflict of interest changing priorities, and data confidentiality. This may lead to a higher rate of denial of access to private lands. In addition, the need for short-term information for regulation might compromise the ability of the agency to commit resources to long-term monitoring programs.
- The agency needs to make EMAP one of its highest priorities internally and in its presentations to Congress. Large fluctuations in funding will seriously damage a program such as EMAP.
- The agency should have a strong administrative and scientific team capable of providing the initiative and scientific leadership required for such a large and highly visible program. The agency should make the commitment of long-term as well as rotating positions for key leadership and scientific advisory personnel.
- The agency should have strong familiarity with each of the resource types being monitored.
- The agency should have the capability to carry out strong research programs
affiliated with EMAP to answer detailed questions raised by EMAP data.

- The agency should have a strong scientific reputation, making it easier to attract top scientists to EMAP.
- The agency should be in close communication with agencies that will administer policy derived from data collected.

EPA clearly has several of these characteristics. EPA scientists conceived the concept of EMAP and were successful in initiating and implementing its predecessor, the National Surface Waters Survey. Certainly, EPA will have some regulatory responsibility in enforcing policy derived from EMAP data, and lines of communication between monitoring and regulatory personnel should be strong within a single agency. EPA in its regional offices has started the promising Regional Environmental Monitoring Assessment Program to address specific questions using an EMAP-like approach. Close ties between REMAP and EMAP personnel will be mutually beneficial.

On the other hand, EPA does not have some of the ideal characteristics. For some of this there may be little EPA can do. For example, EPA is a regulatory agency. That regulatory role was mandated by Congress. Although EPA cannot resolve all the conflicts associated with its regulatory role, it can take steps to reduce their impact, as other agencies have done, by separating research and monitoring functions as much as possible from regulatory ones. There are several other areas where EPA could act to improve the administrative setting of EMAP. As much as possible, funding for EMAP needs to be long-term and predictable. Fluctuations or uncertainty in funding levels will be seriously damaging to a monitoring program this large and ambitious. Within the constraints imposed by the congressional funding process, EPA must act to ensure the institutional commitment to this long-term monitoring effort. EPA should find some mechanism to allow flexibility in timely hiring of qualified personnel at all levels of EMAP. It appears that difficulty in hiring and keeping personnel in key positions has hindered EMAP's progress. While some of the difficulties in hiring personnel are beyond EPA's control, high enough priority may not have been given within the agency to filling key positions, such as the Indicator Coordinator. Finding ways to attract and keep top scientists in EMAP will

Overall Assessment

strengthen the program overall and lead to an enhanced scientific reputation.

3

Program-Wide Components

EMAP's components can be characterized as being program-wide or pertaining primarily to one or more resource groups. This chapter reviews the program-wide components of EMAP—landscape characterization, indicators, and information management.

LANDSCAPE CHARACTERIZATION AND ECOLOGY

Overview

The Landscape Monitoring and Assessment Research Plan - 1994 (EPA, 1994a) sets forth a plan for assessing status and trends of landscape patterns using remote sensing and geographic information system methods. The plan proposes three steps: establishing a baseline condition, detecting changes and determining when and where declines in landscape condition are sensed, and assessing the association between landscape condition and stressors. Such a monitoring effort at regional and national scales is a valuable and crucial component of EMAP.

The Landscape Characterization Plan has not been available for review. The Chesapeake Bay Watershed Pilot Program suggests that the current Landscape Characterization program focuses on land-cover mapping for all resources using thematic mapper data, apparently fulfilling the Landscape Ecology program's goal of establishing a baseline condition. Since the Landscape Ecology and Landscape Characterization programs appear to be inextricably linked, they are evaluated together.

The assessment of land use and land cover is an extremely important activity for EMAP. A large body of current literature (e.g., Houghton et al., 1983; Turner et al., 1990; McDonnell and Pickett, 1993) suggests that human land-use management practices are the most important factor influencing ecosystem structure and functioning at local, regional, and global scales. Land use can dramatically alter species composition, food-web structure, ecosystem carbon storage, and interactions between biota and the atmosphere. Monitoring of the spatial distribution of land cover will provide crucial information regarding our national environmental status.

The Place of Landscape Programs Within EMAP

The role of the landscape programs has changed since the initial phases of EMAP in which Landscape Characterization was described as a central and pivotal program that would provide necessary information for resource groups. With the help of EMAP-Landscape Characterization, resource groups would be able to determine spatial and temporal sampling resolution and key indicator variables. The current role of the landscape programs within EMAP is unclear. A section within the EMAP-Landscape Ecology plan called "Integration with EMAP Resource Groups" only creates more confusion. For example, the most explicit statement is "In this example, a series of common habitat measures could be implemented among EMAP resource groups... ." How will this series be implemented? By top-down decisions? If so from whom? How will information be exchanged? Exactly what personnel, data, concepts, and technology will be exchanged? Although Landscape Characterization is a program with strong potential, without a document describing its goals and structure it is unclear how the program will relate to the rest of EMAP.

Although it is unclear that EMAP-Landscape Characterization is providing integration among all of the resource groups, it has interacted successfully with at least some of them. For example, the EMAP-Landscape Characterization group has worked with the Surface Waters resource group and has characterized the land cover in the watersheds surrounding each of the lakes sampled during the Surface Waters pilot study. These data will be essential in the Surface Waters group's attempt to understand the influence of the surrounding landscape on trophic status, biological diversity, and fishability defined by EMAP as: (the catchability and edibility of fish by humans and wildlife). Clearly, strong interactions between the Landscape Characterization group and each of the individual resource groups will enhance the groups' efforts to understand landscape-level processes that might affect individual resources.

Evaluation

The EMAP-Landscape Ecology plan is extremely well written, providing a coherent theoretical framework for assessing ecological status and trends at landscape scales. This document arguably has provided the clearest description of the connections among societal values, assessment questions, conceptual models of ecological phenomena, and indicator variables. Such a framework is crucial to the implementation of EMAP.

The following are concerns about the landscape programs.

- *EMAP-Landscape Characterization and EMAP-Landscape Ecology are in very early stages relative to the resource groups, with an implementation plan to be developed in the next three years.* It is difficult to see how the landscape programs can guide resource groups if they are so far behind in development.

Program-Wide Components

- *The major impetus, ideas, and follow-through of the Landscape Ecology program appear to rely upon scientists who are not a permanent part of EMAP.* Clearly, it is ideal that one of the leaders in landscape ecology, R. V. O'Neill, along with Denise Shaw of EPA, is guiding the development of the program. But it is unclear whether any permanent personnel or resources, other than the hard-working Denise Shaw, are dedicated to the program.
- The focus of the EMAP-Landscape Ecology program is abstract. A major focus on theory is not in itself a concern. However, there should be a concurrent effort in EMAP to assess ecological functioning. Such an effort would require interaction and integration among resource groups at the empirical stage, relying on common sites and common indicator variables (see Chapter 2).

EMAP-Landscape Ecology's current plans are to study the indicators of landscape structure discussed in the large body of theoretical work on landscape ecology. However, the relationship between these indicators and ecological functioning has yet to be demonstrated. Since individual resource groups focus on assessment end points, the Landscape Ecology program should emphasize ecological functioning.

- *Landscape Characterization is not a well-defined program.* This program is costing millions of dollars, and has the potential to be one of the most important synthesis and integration programs; yet there is no documentation of its purpose, its projected scope, or its connection to the resource-group activities. Questions about the Landscape Characterization program's direction abound. Will thematic mapper data be used to classify the entire United States? Will other satellite imagery be used to assess the nation's status on an annual basis? What will be the result of the collaboration between EMAP and the United States Geological Survey's EROS Data Center? The data center could provide EMAP access to biweekly composite images of the United States at 1-km scales. Such images could provide a useful tool for assessing changes in the nation's ecological status. Will EMAP direct any resources toward this end, or toward combining these data with field data from the resource groups?

There appears to be an important gap between Landscape Ecology and Landscape Characterization in the area of geographic analysis. Large-scale geographic and temporal trends can be extremely useful for testing correlations, such as those between lake and terrestrial productivity, or atmospheric deposition and ecological indices. These correlations can provide strong inference for cause and effect relationships. Further, data from the EMAP sampling sites can be reasonably extrapolated with good geographic information and ecological algorithms. The committee suggests that such analysis belongs in landscape programs.

Providing landscape-level data surrounding each of the EMAP study sites seems to be a major reason for the existence of EMAP-Landscape Characterization. The benefits of the interactions between the individual resource groups and EMAP-Landscape Characterization will be diminished without a strong information management system that will allow the individual resource groups easy access to the land-cover data. Therefore, EMAP needs to develop and maintain an administrative structure that demands close communication and interaction among EMAP-Landscape Characterization, EMAP-Information Management System, and each of the resource groups. This structure might take several forms; one possibility is to locate leaders from each of these groups in a central office.

EMAP INDICATOR DEVELOPMENT STRATEGY

Introduction

A fundamental premise of the EMAP program is that the status of large and complex ecological systems can be monitored and assessed using a limited set of indicators, and biological indicators are being proposed as the key indicators within EMAP (see Chapter 2). Choosing appropriate indicators has been a

Program-Wide Components 51

major focus of EMAP activity since the program began. Despite the obvious centrality of indicator development to EMAP, the completion of a comprehensive indicator-strategy document has been slow in coming. An early version of a strategy document (Olsen, 1992) followed a major program reorganization. A new indicator-strategy document has since been developed and was distributed in the spring of 1994 (Barber, 1994). This more recent document serves as the formal basis for the review of the overall indicator strategy of the EMAP program. This review also focuses some attention on the individual resource groups, since they develop indicators independently of one another.

The EMAP Strategy

The EMAP program has developed a four-part strategy for selecting the indicators it will use for nationwide ecological assessments. These four major steps are *selection*, *evaluation*, *implementation*, and *re-evaluation*. Each step has an associated list of tasks that together constitute the overall EMAP indicator strategy (Table 3-1). This strategy provides a succinct and useful starting framework for standardizing indicator development across resource groups. Earlier indicator documents (Hunsaker and Carpenter, 1990; Olsen, 1992) provided additional background information on EMAP indicator development, much of which has been refined and summarized in the current strategy document. The strategy document strongly and correctly emphasizes the iterative nature of indicator development. It makes clear the desirability of ongoing interactions between EMAP and the administrators, politicians, scientists, and the general public who use the assessment information, as well as the continual need to reassess and redevelop appropriate indicators. The strategy represents a reasonable and sequential procedure that could serve to coordinate what are at present a very diverse set of EMAP indicator development activities.

Table 3-1 EMAP INDICATOR STRATEGY; summarized from pages 17-18 of the indicator strategy document

Stage

Indicator selection
    identify environmental values of the resource
    formulate assessment questions
    identify major stressors
    develop conceptual models of structure and response to stressors
    select indicators for research and evaluation

Indicator evaluation
    evaluate logistics at regional and national scales
    characterize temporal and spatial variability
    develop nominal-subnominal criteria
    prepare example statistical summaries and assessments
    determine sampling densities required to meet data-quality objectives
    select core indicators for implementation

Indicator implementation
    monitor core indicators nationally
    prepare annual summaries
    prepare periodic assessments

Indicator Re-evaluation
    periodically re-evaluate core performance
    identify emerging assessment questions
    conduct research on new indicators

Source: Barber, 1994

Evaluation of the Indicator Development Strategy

The Document. The current strategy document itself is a clear improvement on the original, and shows a significant sharpening and development of thought within EMAP with regard to biological indicators. A question remains, however, as to whether the current document is comprehensive enough to guide such a key element of the overall EMAP program. While it clearly presents an overall strategy for indicator development, it is still weak in terms of clarifying specific procedures and organizational arrangements that will ensure that program development is consistent with the vision of the strategy document.

It is unfortunate that five years into the development of the EMAP program there is little evidence of a clear program-wide set of procedures that standardize and coordinate indicator development. The fact that it has taken this long to produce the current indicator development strategy document reflects poorly on a program that claims to rest on innovative indicator development. The current strategy document can provide a reasonable starting point for such coordination. Because a considerable amount of indicator development work has already been done in the pilot studies of various resource groups, it is imperative that a central strategy and set of guidelines be available.

The Strategy in Theory and Practice. There are already substantial inconsistencies between the EMAP strategy and the actual indicator development taking place in the resource groups. For example, few if any of the indicators now being evaluated in the field have their links with assessment end points and potential stressors documented in scientifically defensible conceptual models, as specified in the development strategy. EMAP established an Office of Indicator Development and appointed an indicator coordinator at EMAP Center early in 1993. However, as of March 1994, the Indicator Development Strategy Document (Barber, 1994) makes no mention of this central coordinating office, the coordinator, or their roles in indicator development. Similarly, the status of an indicator development database, proposed by Olsen

(1992) to facilitate communication among resource groups during indicator development, has apparently not changed in two years. Barber (1994) reported that the database is still in the development stage. The divergence of theory and practice in this matter will only grow without strong central leadership. Individual resource groups have taken very different approaches to evaluation studies and appear to have different strategies and criteria for evaluating the performance of potential indicators (NRC, 1992; NRC, 1994a; NRC, 1994b). The conclusion is that there is still little operational coordination of indicator development within EMAP. However, the committee recently learned that the position of indicator coordinator has now been filled at EPA, which may lead to a more acceptable level of coordination.

Monitoring Philosophy. The EMAP approach to monitoring involves the identification by each resource group of sets of indicators to be used on a national scale. In most cases, these sets of indicators do not yet exist. This is therefore a good time to impose program-wide procedures for indicator selection and development. According to the strategy documents, the following two factors should constrain the selection and development process.

First, the process must be consistent with the EMAP Assessment Framework (Thornton, et al., 1994), which dictates that assessment questions be based on resource values. Resource values give rise to assessment questions; in EMAP these usually focus on resource status with respect to that value.

Second, the selection process should be science-based. That is, based upon current scientific theory and models of resource structure and functioning.

The indicator development strategy clearly calls for the use of explicit conceptual models as a basis for selecting potential indicators for field evaluation that are scientifically sound. These models will provide a scientifically defensible hypothesis showing how indicators are related to resource values and assessment end points. They will also show how indicators are linked to potential

Program-Wide Components 55

stressors and other important aspects of ecosystems structure and functioning.

What is at issue here is the entire EMAP retrospective assessment approach (see Chapter 2), which draws its conclusions from arguments based on the weight of evidence. Documenting the mechanistic link between an indicator and its stressors and assessment end point is crucial to the retrospective assessment approach (Thornton et al., 1994). The problem is, while the assessment framework document discusses the issue of establishing cause and effect in some detail, the indicator strategy document says little on this key point. EMAP's retrospective approach will not succeed unless the development of conceptual models is taken seriously.

Put more simply, it is at this stage that science should be infused into the EMAP assessment protocols. Unfortunately, the example models used in both the strategy development and the assessment framework documents (and for the most part in the resource group planning documents reviewed in previous reports) are almost uniformly trivial. This raises questions as to whether EMAP is capable of implementing its own assessment strategy, and whether it is interested in pursuing indicator development in a scientifically rigorous fashion. In many cases EMAP seems to confuse conceptual ecological models with graphic portrayals of rationales for indicator selection decisions that have already been made. Each EMAP resource group should develop one or more *mechanistic* conceptual models of its resource based on current scientific knowledge. These models should serve as explicit hypotheses about those aspects of ecosystem structure and functioning relevant to the assessment end points the group has chosen. The models must be detailed enough to include potential indicators, assessment end points, key variables, and factors causing the endpoint to vary across the landscape. The models must also include some hypothesis concerning mechanistic and functional relationships among key variables. Only from such models can the resource groups generate lists of potential indicators about which there is sufficient knowledge to interpret changes in indicator status. It is necessary to link indicator selection to current understanding of ecological mechanisms if indicator development is going to be scientifically credible, and if the resulting monitoring data are to be useful in ecological assessment.

Evaluation of Indicators. The strategy document seems particularly weak in terms of laying out guidelines for indicator evaluation. Developing indicators of ecological condition as envisioned by EMAP is a challenging task. It is uncertain how much the anticipated indicators of ecological condition can be developed, and how reliable and easy to measure they will be. Therefore, to be successful, indicator development for EMAP will need to have a substantial research component.

There are many practical, program-wide questions that need to be answered. How do resource groups decide that an indicator works? What kinds of evidence are required? How does the program review the evidence? How often and in how many different settings should validation tests be performed? How many indicator variables should be measured for a given assessment endpoint? These are just some examples of issues in indicator evaluation that need to be answered by all of EMAP in a coordinated manner. The current indicator development strategy provides insufficient guidance in these matters. The document does contain a list of what are identified as criteria (see Barber, 1994, Table 2, page 37); but these really seem to be only a list of desirable general characteristics. A feature of a useful criterion is that it can be (at least relatively) unambiguously evaluated. The current document provides little information as to how EMAP will evaluate the performance of a potential indicator with respect to Barber's Table 2, nor does it discuss whether any or all of the criteria are essential.

There are numerous evaluation issues that need program-level guidance if the indicator selection strategy is going to yield the nationally applicable set of indicators EMAP envisions for each resource group. One particularly important issue, for example, is the explicit and early evaluation of the biological and statistical properties of all potential indicators. This should be given a high priority. Such evaluations should include analyses of the properties of the mean, variance, and behavior of the index in power tests. (A power test is a test of the statistical power of an approach to detect change). If this cannot be done analytically, then simulation analyses should be performed. An example of such an investigation for Karr's Index of Biotic Integrity is that done by Fore et al. (1994).

Questions of Scale and Regionalization. The indicator strategy document wisely provides for a phased indicator evaluation (Table 3; Barber, 1994, page 41). Problems associated with spatial scale in indicator performance are likely to be common, as noted above, and careful testing in pilot and then demonstration projects before national implementation is likely to be the most efficient approach in the long run. Biological indicators useful at a national scale at present are quite rare. Biological indicators generally require region-specific interpretations. For example, green sunfish (*Lepomis cyanellus*) play an important role as in indicator of subnominal habitat for states of the central midwest. This species does not occur in colder rivers of the northern midwest, but white suckers (*Catostomus commersoni*) may provide a useful analog. Similarly, sweet gale (*Myrica gale*) is a common indicator of rheotrophic wetlands in the upper midwest, but is a common constituent of coastal ombrotrophic bogs in New England. If EMAP succeeds in developing a national set of indicators, this will be a major accomplishment indeed. EMAP's emphasis on indices and cumulative distribution function hopefully will compensate for large-scale spatial variations in an indicator's effectiveness. Karr's Index of Biotic Integrity (IBI) is an example of an index that might be used on a national scale, but it has to be adjusted from region to region. However, careful evaluation of each potential indicator over larger and larger areas and regions seems a wise approach. The ways in which the various resource groups deal with this problem will have important consequences for the selection of nationally implemented indicator metrics. Program-wide strategies for dealing with this issue should be developed now, in time to be applied with some uniformity across the resource groups.

Indicator Development Across Resource Groups. This is another area in which more coordination is needed. Ideally, nationally-implemented indicators should have some correspondence across resource groups. At present, it is unclear whether or not the assessment questions in each resource group are similar enough to lead to parallel sets of indicators. Possible examples include indicators reflecting net primary productivity, biological diversity, and aesthetic value. Such symmetry among resource groups, while not essential to the basic EMAP objectives, would greatly enhance the scientific and analytical value of the data collected.

Summary

The indicator strategy (Barber, 1994) is a welcome if late addition to the EMAP program. It is clearly sufficient to begin the coordination that will be required to bring the indicator development activities of the resource groups into a more scientifically rigorous context. Having a documented strategy, however, is not a replacement for organizational structure and guidance in this area. The strategy document, in conjunction with a strong central office for indicator coordination, would be a key asset in the EMAP program, and would help to ensure that the massive amounts of data that EMAP proposes to collect will in fact be useful in future retrospective assessments. Without such coordination, the ad hoc nature of indicator development as it currently operates in the resource groups will weaken the value of whatever data the monitoring program generates.

An encouraging sign is that the position of indicator coordinator has recently been filled. This can be the first step in a more rigorous coordination of indicator development within EMAP. The peer-review committees have been essential in providing a sounding board for the various resource groups over the past five years, but it is clear the relationship between EMAP and these boards is too informal to provide the degree of coordination required. These boards have no formal oversight authority, and operate in isolation from each other and the overall EMAP effort. The role of peer-review panels as currently constituted in the program should be expanded, but they are no substitute for much needed central coordination of indicator development inside EMAP itself.

INFORMATION SYSTEM

The EMAP-Information System is a critical component of EMAP. It is essential for the success of the program that EMAP-Information System provide appropriate access to the data and information generated by the program, as well as the personnel, hardware and software resources necessary to support such access and processing.

EMAP-Information System should be viewed primarily as a scientific database system that also supports data analysis and modeling activities. The requirements for systems that support the storage of, access to, and analysis of large-scale collections of scientific information are not satisfied by the relational database systems currently available. Therefore, although these latter systems have proven to be the technology of choice for supporting many business activities, they have not proven adequate for supporting scientific activities, particularly modeling efforts. Scientific data typically involve very long transactions, are transformed for a variety of simulation processes and models, have heavy reliance on metadata, and are more complex in structure and organization. As a result of these fundamental differences, the EMAP-Information System must be carefully designed to facilitate the myriad future uses appropriate for scientific data that could easily be overlooked when trying to retrofit existing relational database technology for purposes for which it was not designed.

A scientific information system can be viewed as having two logically distinct components. The first component is the environment in which the users will operate and the second is the technical personnel, hardware and software to support this environment. The environment must be designed to support the users in their information-processing tasks and permit users to carry out their tasks efficiently. There must be efficient support for database activities, including data input, data access, and data analysis. EMAP-Information System user requirements involve databases containing a large and heterogeneous collection of data sets, which are spatially distributed and spatially indexed. In relation to such a database, EMAP-Information System users should also have access to a large variety of analytical and visualization tools that may be applied to various subsets of the data, over various spatial scales, and at various stages of information processing.

EMAP-Information System users should also be able to integrate any tools they need into their environment. An investment in technical support staff and good hardware and software should ensure that the information management system enhances the productivity of the users and facilitates the accomplishment of EMAP's goals.

The information available to the committee for its evaluation of EMAP-Information System was a strategic plan for the information system and an initial rapid prototype system developed largely within the estuaries resource group. A set of white papers intended to lay out important details of the system and its design have not appeared. There is, however, a rapidly growing body of information concerning scientific database systems that provides standards for comparison. In particular, much information has been generated in relation to programs such as the Long-Term Ecological Research and National Science Foundation scientific database initiatives, and National Aeronautics and Space Administration programs such as the Applied Information Systems Program and the Earth Observing System project.

Concerns About EMAP-Information System

The list of serious concerns that follow are based on a visit to EMAP headquarters, discussions with EMAP-Information System staff, and the limited documentation available.

• The high level of abstraction in the strategic plan does not permit an adequate evaluation of the proposed system. In particular, many elements that one would expect to find in a strategic plan, such as a requirement analysis, system design, and other detailed studies and plans, are absent. Much of the effort underlying the document appears to have been focused on developing high-level concepts about the functionality of the system, its implementation, and its management in terms of the "Zachman Framework". This framework, which was developed within EPA, takes a management point of view towards system specification. While a management perspective should not be ignored, such planning should be based on the viewpoint that EMAP-Information System is a scientific database system, rather than an information system focused on the needs of management. In particular, the planning should be largely focused on the design of an environment that is sensitive and responsive to user requirements and on the design of appropriate hardware, software and personnel for such an environment. Furthermore, the planning should follow a top-down strategy that begins with the highest-level requirements of scientific users, and gradually expands the details of these requirements.

• The strategic plan lacks almost any reference to recent activities in the area of scientific databases and supporting infrastructure. In particular, the plan makes a single reference to an early conference on scientific databases. There is no reference to database developments in programs that have similarities to EMAP, such as NSF's Long-Term Ecological Research program. There is no reference to emerging protocols and standards, such as the recent federal standard concerning spatially referenced data. There is no useful reference to the role in the EMAP-Information System that the Internet, World Wide Web, and other related developments will play, although these are affecting scientific activities and scientific database systems in important ways. It appears that EMAP-Information System is being developed in an environment that is effectively isolated from related developments occurring outside of EPA, despite the extensive activity in the area of scientific databases. It is not clear whether EMAP drew upon appropriate expertise in drafting the strategic plan.

• The plan gives no significant indication that user requirements, and their central role in EMAP-Information System, have been given adequate consideration. In particular, the plan does not include any useful analysis of user requirements. An appropriate basis for a strategic plan is a detailed analysis of the user requirements that are likely to arise in EMAP over the next two decades.

• The plan fails to specify important aspects of the user environment that relate, for example, to data access, data processing, and data visualization. Furthermore, the plan does not appear to provide a uniform approach over the resource groups with respect to such issues. Currently it appears that the plan allows much autonomy among the resource groups with respect to important decisions concerning issues such as query languages, data models, and data organization. Uniformity over the groups with respect to matters such as these may well be critical for the success of EMAP, since long-term coordination of both data and analyses of data across resource groups will be not only important but essential to achieve the degree of integration envisioned for EMAP.

• The plan fails to address the specifics of such complex and intensive data processing. For example, EMAP's success relies on the handling and integration of large, spatially indexed data sets, including images, with more standard observational data sets. The failure of the plan to specify important aspects of support for the users, such as technical support staff, hardware and software, may result from the plan's failure to describe the environment.

Program-Wide Components

- The plan does not specify how the development and operation of EMAP-Information System will relate to other federal programs in terms of sharing resources and technologies (for example, making EMAP data available to the National Science Foundation Center for Ecological Analysis and Synthesis). While there is some interaction with a consortium of federal agencies involving a large, spatially indexed database located at the EROS Data Center, the links between such systems and EMAP-Information System are at best vague. They are particular challenges for EMAP-Information System in its ability to stay abreast of developments relating to the national information infrastructure.

- The plan appears to underestimate the resources required to design, construct, and test an information system that will prove adequate for EMAP in the long-term. For example, the plan appears to call for only 1 to 1.5 full-time positions over the next four years to address the issue of spatially indexed data. With such inadequate resources it would be a surprise if an adequate or stable system were available to EMAP users by 1997.

- It is not clear that the rapid prototyping effort has a clear-cut goal. Such a system should be focused on understanding user requirements. The results of this understanding could then be incorporated into the main EMAP-Information System. The current prototyping effort, which involves an integration of data management systems such as ORACLE, ARC-INFO, and SAS, has very limited capabilities and is inefficient from a user's point of view. The system appears to be focused on resolving technical issues with the use of systems that are probably not appropriate for supporting large-scale scientific database activities in the long run.

Summary

The information provided for this review of EMAP-Information System is an insufficient basis for evaluating the adequacy of the system in its primary task of supporting the resource-group scientists. In particular, too many aspects of the system are specified at too high a level of abstraction for an appropriate judgment while many important aspects either are left unspecified or are severely underspecified. Based on our reading of the strategic plan and on our evaluation of the rapid prototype, the committee has very serious doubts as to whether the current approach to designing, implementing, and managing EMAP-Information System is an appropriate solution to the long-term data and information processing requirements of EMAP.

4

Resource Components

This chapter considers the various EMAP resources. In some cases, resources have been reviewed in earlier reports on surface waters, estuaries, and forests; for those, the executive summaries of the earlier reports are reproduced along with any new information. Brief, new reviews are provided here of the agroecosystem and Great Lakes. One of the program's originally planned resource groups, wetlands, has been eliminated, and insufficient information is available on the rangelands (formerly arid lands) resource group for review.

AGROECOSYSTEMS

The activities of the Agroecosystems Resource Group of EMAP are reported in five documents: (1) *Agroecosystem Research Plan*, Heck et al., 1991a; (2) *Agroecosystem Monitoring and Research Strategy*, Heck et al., 1991b; (3) *Agroecosystems 1992 Pilot Project Plan*, Heck et al., 1992; (4) *Agroecosystem Pilot Field Program - 1993*, Campbell et al., 1993; and (5) *Agroecosystem Pilot Field Program Report - 1992*, Campbell et al., 1994. The monitoring phase of the Agroecosystem component of EMAP is scheduled to begin in 1998. To date the program has been largely concerned with research on biological indicators, establishment of relationships with the National Agricultural Statistical Survey of the United States Department of Agriculture that will collaborate with EMAP and collect the majority of the field data, and refinement of the logistics of data generation from sample collection to data analysis and reporting.

The objectives of the Agroecosystem component of EMAP are the same as those of the other resource components, i.e., the determination of the distribution and extent of agroecosystems, assessment of status and trends, association between changes in status and stressors, and preparation of periodic reports. While this consistency is necessary and appropriate, the application of the EMAP approach and protocol to highly managed agricultural ecosystems is problematic. Nearly all past agricultural research has focused on management of agricultural commodities. Practically nothing is known about the other biological components of agroecosystems. Gathering such information is problematic as the intensive management of agroecosystems can overwhelm measures that would be meaningful in other ecosystems. For example, the application of nitrogen fertilizer causes extreme short-term changes in soil chemistry and microbial community structure and metabolism. It is difficult to avoid these effects completely in a sampling protocol because nitrogen fertilization practices vary with crop, agroclimatic zone and individual farmers. Development of meaningful indicators that are relatively free of influence by agricultural managers is an extraordinary challenge, but it is necessary given the frequency of sample collection.

The strategy of the Agroecosystem component involves monitoring of indicators that are clustered around three assessment questions and collection of samples and data by the National Agricultural Statistical Survey. The assessment questions are expressed differently in several of the documents produced by the resource group but are reasonably consistent in intent. They include:

- sustainability of production potential for commodities;
- quality of air, water and soil; and
- maintenance of biological diversity.

A fourth and implied assessment question involves the impact of agroecosystems and their management on adjacent and downstream ecosystems. These assessment questions seem appropriate and comprehensive. However, the development of quantifiable indicators for such questions is difficult.

The measurement of physical and chemical properties of air, water, and soil and the identification of trends in such measurements are straightforward. However, integration of these properties into a measure of quality and establishment of acceptable levels of quality is far from straightforward. The assessment of the sustainability of production potential is limited by the current state of the art to monitoring of crop yields. The relationship of yield to ecosystem condition is unclear. During the past decade, yields of wheat and rice agroecosystems in southern Asia have declined. Whether or not this decline represents a compromise of the sustainability of production is as yet undetermined, despite considerable research efforts to identify the cause of such declines. EMAP's attempts to assess the value of production efficiency as a measure of sustainability are admirable, but they are still in the research stage.

Because yield data are already available, EMAP's greatest contribution to monitoring agroecosystems is the development of indicators of overall ecosystem status. Appropriate indicators of biodiversity within agroecosystems are not yet perfected. EMAP's exploration of the potential of using trophic groups of nematodes as indicators of soil health and diversity is founded on good theoretical grounds, but experimental testing of the theory is still under way. The recent addition of hedgerows and pest-management parameters will expand the number of habitats sampled, and thereby likely improve overall estimates of system diversity. The challenge for EMAP is to ascertain the set of taxonomic divisions and habitats that will reflect the system diversity.

Because of the regional variability in the nature and distribution of agroecosystems, EMAP should test the adequacy of the sampling grid. In the Southeast and Midwest, agroecosystems are relatively uniform in distribution because the landscape is relatively uniform, and rainfall is generally adequate for growing crops. The base grid will likely sample such regions very well. In the arid regions of the West, and especially in central California, agroecosystem distribution is very patchy because agroecosystems are clumped around sources of irrigation water. High-volume crops such as fruits and vegetables are generally grown in these areas with high yield and economic return. In the Northeast, agroecosystems are typically patchy because of the topography. Agroecosystems with patchy distribution are less likely to be sampled adequately by the base grid.

EMAP intends to collaborate with the National Agricultural Statistical Survey in the collection of data. Essentially, EMAP will augment the sampling grid and data requirements of the survey, which has been ongoing for years. The partnership will capitalize on the experience and expertise of the United States Department of Agriculture while contributing the ecological experience and perspective of EPA and EMAP. The ecosystem components that are not commodities and the influences of agroecosystems on adjacent and downstream ecosystems are concerns that EMAP will add to the survey. EMAP should be commended for forging this partnership to increase the utility and cost effectiveness of the efforts of both agencies.

The 1992 pilot program of the EMAP Agroecosystem group was conducted in North Carolina by a new partnership between EMAP and the Department of Agriculture National Agricultural Statistical Service (NASS). This partnership represents an attempt to modify the NASS survey of U.S. agricultural lands to be compatible with the sampling and data needs of EMAP. The pilot was conducted as the first attempt to perform field operations, collect and prepare samples and field data, transport samples to various laboratories for analysis, and to manage and analyze resulting data. In essence it was a true operational pilot study.

In most respects the pilot was a qualified success. Data collection was incomplete for several reasons, including failures by the National Agricultural Statistical Service (NASS) to adequately adapt, and problems in quality control. None of these failures is surprising in a first trial. However, it is essential to resolve these issues before implementing the EMAP monitoring program across a greater area.

The pilot failed to establish the suitability of its indices and measures as indicators of ecological condition. While some of the indices and measures have potential, their values are not yet documented. The report of the pilot recognized its limitations and failures, but few details were offered about correcting the deficiencies in future activities. It is unlikely that the lessons of the 1992 pilot were used to modify the 1993 pilot, as the report of the 1992 study was published in 1994, well after the second pilot was completed.

The Pilot Field Program of 1993 was conducted in Nebraska. It had four major objectives as follows: (1) empirically establish the range and variance in indicator values within the State of Nebraska; (2) to compare the efficiency and precision of the EMAP hexagonal design and the NASS rotational panel design; (3) refine plans for logistics and data handling; and (4) develop and evaluate additional indicators of soil quality and landscape structure. Also included within the first objective was a subobjective to assess the ability of indicators to reflect condition. These are generally worthwhile objectives, and the plans for implementing them seem sound.

Only the subobjective of assessing the ability of an indicator to reflect ecosystem condition seems questionable. The establishment of the correlation between indicators and ecosystem condition is a great challenge faced by EMAP across all resource groups. It is particularly challenging to the Agroecosystem group because there is so little ecological knowledge available on agroecosystems. The challenge will only be met by persistent research and empirical evaluation of indicators.

Conclusion

The Agroecosystem resource group is subject to most of the same concerns that have been expressed about other resource groups. The utility of indicators, especially the biological indicators, is largely undocumented and requires empirical research over several years to provide such documentation. The intensity of the base grid may be too coarse to adequately sample adequately agroecosystems that have patchy distributions. Questions concerning data management, coordination with other resource groups, and interagency cooperation are virtually identical to those in other resource groups. Most problematic is the appropriateness of the EMAP approach, primarily intended for natural systems, for monitoring intensively managed agroecosystems.

The Agroecosystem program is at an immature stage of development relative to other resource programs. This is not surprising as the focus of past agricultural research has been on management of a few species and particularly on factors affecting their yields. Ecological sustainability, and its relationship to biotic components, and the functional linkages among such components are relatively new concerns for to agricultural scientists. Efforts are under way at several institutions to develop methods for assessing agricultural sustainability. EMAP would be well served to become familiar with such efforts and to incorporate them in its deliberations and plans. Examples include the Sustainable Agriculture Research and Extension-Agriculture Compatible with Environment (SARE-ACE) and Sustainable Agriculture and Natural Resource Management (SANREM) programs at the University of Georgia, work at the Leopold Center of Iowa State University, and the Agroecology program at the University of California at Santa Cruz.

ESTUARIES
(Modified from NRC, 1994a, Executive Summary)

The goals of the 1990 Virginian Province demonstration project were to identify which indicators and design attributes are most effective for assessing the ecological condition of estuarine resources on a regional scale with limited financial resources. Significant progress was made in many areas.

The grid-sampling scheme was successfully modified to represent better discrete systems such as small estuaries and large rivers without compromising the acquisition of unbiased samples. A complex field-sampling program was successfully mounted with a well-coordinated plan for quality assurance of data acquisition, analysis, and management. Initial steps were taken toward the development of a group of indicators of ecological condition. Subsequent efforts have been made to involve regional managers by having them cooperate in future sampling and in the evaluation of the applicability of data collected. The activities and results of the first year of sampling (1990) have been issued in a well-written synthesis report describing the process of indicator development and containing an initial interpretation of the data obtained (Weisberg, et al., 1992). Based on the material in this report, EMAP has made a good first step in getting the estuaries section of EMAP started.

Although there have been many positive accomplishments, there are a number of areas needing significant work. A more explicit conceptual model must be developed to drive indicator development and set priorities. Continued work also is needed to develop meaningful indicators that assess basic ecological condition (status and functioning).

The combination of the EMAP probability-based sampling design and the realities of national coverage with a limited budget severely limit the type and number of indicator measurements that can be made. The review panel of the Estuarine Research Federation doubts that the indices generated by EMAP will have the power to detect the amount of environmental change expected. Environmental change can occur at various rates. For example, the EMAP design standard is the ability to detect a 20 percent change occurring over a decade. The published information on changes in various indicators shows, however, that some changes occur in estuaries at a much slower rate than this (Stanley, 1993). As a result, it may take several decades for a 20 percent change to occur, therefore, it would take decades to be able to detect changes with the current EMAP sampling design. By contrast, changes in ecosystems can be quite sudden and catastrophic, perhaps too fast to be adequately captured by EMAP's sampling scheme. It is time for this issue to be clearly analyzed by EMAP using extant data sets or similar proxy data. According to the letter from Dr. Martinko (Appendix A), these studies are currently being carried out on Virginian Province data sets. Future (1995-1996) analyses are planned on existing data sets from the Louisianian Province.

Large programs such as the estuaries component of EMAP usually pay insufficient attention to analyzing exactly what they have learned in their pilot and demonstration projects. There is a temptation to think that the next challenge is to carry out pilot projects on new provinces, one after another. However, as pointed out by the Estuarine Research Federation review committee, the real challenge is in obtaining the best possible set of indicators of ecological condition. *Therefore, EMAP personnel should stop and evaluate the estuaries part of EMAP in detail before going on or adding additional provinces. This evaluation should occur as soon as possible after the Virginian Province demonstration completes its first four-year cycle and should include a comparison of the EMAP information with other published information on indicators of condition of estuarine resources with different design attributes.* This evaluation has begun (See Appendix A).

Resource Components 73

Indicators

1. The estuaries component of EMAP should include indicators of ecosystem function. These indicators are difficult to monitor when studies are made only once a year, but can be estimated to some extent indirectly. Lack of such indicators should be addressed as soon as possible. An example of such an indirect approach is that algal biomass can be used as a surrogate for primary production. Also, remote sensing provides one possibility for chlorophyll measurements on a regional scale.

2. Another measure of important coastal habitat, submerged aquatic vegetation, was missing in the Virginian Province demonstration project. Inclusion of submerged aquatic vegetation in the sampling scheme for all estuaries demonstration projects should be considered.

3. Insufficient effort has been devoted to fish sampling to make the data obtained useful. A relatively short trawl done once does not collect enough fish for meaningful determination of population characteristics, contaminant body burdens, or incidence of gross pathology. If quantitative information on these indices is desirable, arrangements should be made with other agencies with the experience and personnel on hand for more comprehensive collection and analysis of data. If the level of planning and effort allocated by EMAP for these activities cannot be significantly increased, fish sampling program should be eliminated.

4. In support of new indicators development, areas of research that should be looked at include the analysis of long-term data sets from various sites to examine indicator variability and its causes and the use of molecular probes to look for the presence of enzymes indicating pollutant exposure or changes in ecosystem function (e.g., nitrogen fixation).

5. No use has yet been made of a number of historical data sets in the Virginian Province that have data comparable to those being collected by EMAP. Out of 18 studies investigated, eight were found to contain information important to EMAP and in

particular to EMAP-Estuaries. This material should be analyzed in detail to provide valuable information on spatial and temporal variability and the power of certain indicators to detect trends within a given period. It is past time for this work to have been completed. (This recommendation is addressed in Appendix A.

Advice, Consultation, and Scientific Review

EPA has sought advice from a wide variety of scientists in developing EMAP, but the effectiveness of the present mechanisms for incorporating scientific expertise into the design and execution of resource-group activities are not what they should be. Working groups, which have been used by most of the resource groups, provide peer review but are not necessarily efficient or adequate for ensuring that activities are based on the best scientific approach. Continuing oversight and review by groups of scientists from outside of EPA, built into the program at the highest levels, should be implemented for EMAP Center planning, for indicator development strategy, for landscape characterization, and for all resource groups.

Update

There has been progress since the last report dealing with estuaries was written. Some of the progress has been responsive to the report. Much of the recent progress in the development of indicators and sampling methods has not yet appeared in documents available for review, and so the following evaluation of progress may not be complete.

Two big issues face EMAP-Estuaries (and the other resource components)—indicator development, and the time and space scales at which interpretation of the measurements will be meaningful. EMAP-Estuaries appears to have made some progress with respect to indicators, but the information available does not

Resource Components

indicate whether the program will be able to adequately assess trends in the condition of the nation's estuaries.

Indicators

EMAP officials report that a revised conceptual model for indicator development will be released later in 1994. It should be based on ecological relationships and on how disturbances of the ecosystem would be reflected in the various indicators to be measured. For instance, excess nutrients would lead to hypoxia that could be measured by increases in the extent of areas with low dissolved oxygen. Thus far, it is not clear from EMAP's response to the previous report, which called for explicit conceptual models, that this is the type of model being developed.

Detecting Trends

The fish-sampling program is fundamentally flawed. It is not clear how a supportable program could be undertaken without a large increase in investment of resources for adequate development of indicators. The type of extensive sampling that supported the development of an approach for measuring dissolved oxygen would have to support the development of any indicator based on measurements of fish collected.

It is still not clear that the EMAP sampling design and indicators will have the power to detect the kinds of environmental changes anticipated at the appropriate scales. This is because of the high degree of spatial and temporal variability in estuarine systems and because rates of change are expected to be slow. Stanley (1993) was forced to lump data into 10-year increments and into three major river sectors to detect change, even with monthly sampling at many sites for more than a decade. EMAP is primarily concerned with changes in the areal extent of indicator distributions. However, based on the known spatial and temporal heterogeneity, it is probably not possible to predict whether trends in areal extent of indicator distributions will actually be observed over a 10-year period. Complete analysis of the 4-year data set already in hand for the Virginian Province will provide the first real indication of whether the design criteria will be met and how meaningful the information might be.

In their response to the National Research Council review of EMAP-Estuaries, EMAP scientists have indicated their conclusion (see Appendix B: letter from G. Foley to NRC committee, May 9, 1994), based on analysis of the Virginian and Louisianan demonstration projects, that EMAP-Estuaries indicators have the potential to detect a change of 1 to 2 percent per year over a 10-year period. However, there has been no convincing evidence that linear, decadal changes of 1 to 2 percent per year will occur at the scale of standard federal regions.

EMAP has reported plans to devote additional effort to examine the effects of multiple design modifications to EMAP's ability to detect different types of trends. Step function should be included in these examinations. EMAP also has undertaken a four-year assessment of the Virginian Province demonstration project. These are good efforts, and the results should be published as soon as possible. Four years of sampling have now been completed in two estuarine regions: Virginia and Louisiana. Careful analysis of these two efforts should be the cornerstone for further testing of indicators and development in all the estuarine regions. Taking time out from field sampling for careful analysis of information obtained to date is critical to strengthening the program.

FORESTS (Executive Summary, from NRC 1994a)

EMAP's Forest Health Monitoring Program (EMAP-Forests) proposes to collect data on environmental factors that influence forest growth, as well as additional response variables of the trees such as soil nutrients and canopy structure. If this is implemented, the resulting data sets will be valuable for ecologists and foresters seeking to understand basic ecological patterns and for policy makers who require information for the evaluation of future environmental impacts on the nation's forests.

The multi-agency partnership of EPA, the Forest Service, state forestry agencies, the National Park Service, the Fish and Wildlife Service, the Tennessee Valley Authority, and the Bureau of Land Management is exactly the type of cooperation that EPA sees as vital to EMAP's national monitoring effort. Many of the positive features of the program derive from the previously established U.S. Forest Service Health Monitoring Program. What follows is a summary of specific recommendations.

Lack of a Theoretical Basis

Elements of a theoretical basis for EMAP-Forests included hierarchy theory, sampling theory, epidemiological theory, and the stand-development theory of Oliver and Larson (1990). However, the logical basis by which these theories explain the responses of forests to stress, and the subsequent responses of surface waters to changes in forests and by which the theories for indicator development and sampling protocols is not clear. Heavy reliance appears to be placed on a purely epidemiological model. Epidemiological models describe how diseases spread through populations. Such models appear to have little utility in predicting how nutrient cycles, nutrient losses, or biodiversity of ecosystems change in response to stress. Therefore, EMAP personnel should continue development of a theoretical basis for EMAP-Forests from which predictions can be made of general types of forest response to different types of stress. The theory should at least encompass productivity and diversity.

Select a Set of Indicators

It is essential that EMAP-Forests choose sets of indicators with a consistent theoretical basis across all regions as soon as possible and then conduct the staff work necessary to establish sampling methods and convey these to field crews. The next step is to develop the process for interpreting the results derived from field studies. Priority should be given to the evaluation of measurements that integrate limiting factors over the growing season, such as nutrient availability as measured by adsorption in resin bags, and that can be performed quickly using standardized procedures.

Revise Sampling Design

The current design of four-year plot rotations should be replaced or augmented by a design in which some plots are revisited every year. Revisiting a site only once every four years prevents EMAP from making site-specific estimates of changes in these frequencies. This is unfortunate in the case of known cyclical events. An augmented sampling scheme that would permit some plots to be revisited every year would maximize temporal coverage. Some plots could be sampled on a rotating basis to maximize spatial coverage. This is essentially the recommendation also reached by the statistical sampling design team.

Information-Management System

EMAP-Forests should develop a comprehensive information-management plan that outlines user requirements, examines long-term implementation of hardware and software, and fits in with the overall plan for the information-management system.

Publish Study Results

EPA should encourage publication of study results in peer-reviewed science journals to gain credibility in the scientific community and to ensure accessibility of information.

Delay Full Implementation Until Results of Demonstration Projects are Evaluated

For all the reasons described above, the EMAP-Forests program should not be fully implemented until the results of demonstration projects have been thoroughly evaluated and a realistic estimate of the program's costs to EPA and other agencies is available.

GREAT LAKES

The Great Lakes are one of the three main aquatic resource groups within EMAP. In contrast to Surface Waters and Coastal Waters, the Great Lakes were added as a resource group after the program was started. Of the aquatic resource groups, it is the least defined and developed, and relatively little field or pilot-level activity has been directed toward it. EMAP has offered the following documents for review: (1) a draft strategy document prepared for the Great Lakes component of EMAP (EPA, 1992a), (2) a review of that document (EPA, 1992b) and a response to the review (EPA, 1992c); and (3) a status report for pilot field activities conducted in 1992 (EPA, 1994b). This section briefly reviews the nature of the draft strategy document and 1992 pilot studies, comments on the merits and completeness of this work, and raises some questions about the overall Great Lakes program.

The 1992 draft strategy document for EMAP-Great Lakes (EPA, 1992a) is scientifically credible and provides a detailed analysis on several aspects of the proposed Great Lakes program. The authors demonstrated good familiarity with the literature regarding Great Lakes biota in relation to possible indicators to be measured for EMAP-Great Lakes. The report also has clear descriptions of sample design considerations relative to some of the four resource classes in the Great Lakes that EMAP intends to monitor and statistical methods to be used in data analysis. The peer review report on this document (EPA, 1992b) was thorough and thoughtful. It raised several philosophical concerns about the approach described in the strategy document and had many specific technical questions, especially about indicator selection and data analysis. Some of these concerns are similar to those that other reviewers have raised about other components of EMAP. The peer reviewers were especially concerned about limited attention that had been paid to the sampling design for the most variable, productive areas of the Great Lakes, (like harbors, embayments and wetlands. It also expressed concern that the overall EMAP design drives the entire program, and that the design unjustifiably influences the selection of indicators (e.g., only those indicators that can conform to such a design will be selected) (EPA, 1992b). The authors of the draft strategy document responded to each comment and criticism of the peer review panel in a separate report (EPA, 1992c). In general, the responses indicate that the authors understood and considered the comment or criticism, and in many cases they indicated that further studies were planned or underway that would address the issue. Overall, however, it appears that work of the peer review panel—its comments and recommendations—had little influence on the plans of EMAP for its Great Lakes program and that no significant changes were made in the strategy document as a result of the peer review.

According to EPA (1992a) and EPA (1994b), EMAP's objectives for the Great Lakes are essentially the same as those for the overall program and for other resource groups (see Introduction). The primary environmental values EMAP has identified for the Great Lakes (i.e., its assessment end points) are biological integrity and trophic condition. (It should be noted that the Fiscal Year 1992 status report is not entirely consistent with regard to the former end point; the executive summary refers to "biotic integrity," but Section 1 of the text uses the term "biological integrity"; these are not necessarily exactly the same concepts.) The draft strategy document and 1992 status report mention several other end points that are acknowledged as important designated uses for the Great Lakes both in the past and in the present: fishing, swimming, navigation, drinking water supply, and habitat for aquatic life. Furthermore, the reports concede that "not all the societal concerns about the conditions of the Great Lakes fall neatly under the biotic umbrella." Why the program chose to define its interests regarding conditions in the Great Lakes so narrowly is puzzling, and a more comprehensive monitoring and assessment program with a broader array of assessment end points would be more appropriate and better suited to the agency's overall mandates than the narrow program it proposes.

EMAP has delineated four resource classes within the Great Lakes that it regards as amenable to assessment within the EMAP sampling grid: (1) coastal wetlands, (2) harbors and bays, (3) nearshore regions, and (4) offshore regions. The 1992 pilot studies focused on the offshore zone and to a lesser extent on the nearshore zone in the two uppermost lakes, Superior and Michigan. Spring cruises were conducted on both lakes, and two additional cruises (summer and fall) were conducted on Lake Michigan. These cruises were designed to address the following issues:

(1) the adequacy of the base EMAP sampling grid to assess trophic conditions in offshore areas and comparability among trophic state data collected by EMAP (at grid sites and data collected under other Great Lakes sampling programs);

(2) the appropriate index periods in which to assess trophic conditions in the lakes and collect benthic invertebrate indicator organisms;

(3) the definition of nominal conditions for sediment indicators in nearshore areas;

(4) evaluation of the use of diatoms as representatives of Great lakes phytoplankton populations;

(5) investigation of the use of sediment cores for historical trend analysis of diatom populations; and

(6) evaluation of the use of sediment traps to collect integrated samples for estimates of annual diatom populations.

A large number of narrowly defined questions were addressed within the context of the above six issues. In general, analysis of the data appears to have been done in a professional and competent manner. However, the overall significance of some of these analyses in the context of designing and implementing EMAP-Great Lakes is not always clear. A few of the conclusions reached by the authors seem difficult to explain based on data presented in the report, but a detailed evaluation of the basis for conclusions about narrow scientific issues is beyond the scope of this report. In summary, the status report is a satisfactory compendium of the field work undertaken in 1992 and follow-up analyses done on the data, but it does not provide much insight into why the particular work was done (i.e., how it fits into the overall strategy to design a Great Lakes monitoring and assessment program).

The primary sampling data to which the EMAP trophic state data were compared in Lake Michigan are those obtained by EPA's Great Lakes National Program Office (GLNPO) under the Great Lakes International Surveillance Plan (GLISP). This plan is a bilateral arrangement between the United States and Canada under the 1972 Great Lakes Water Quality Agreement, which has produced a long-term record of trophic state and toxic contaminant data for fixed stations at offshore sites in the Great Lakes. In brief, the 1992 pilot sampling program found that EMAP and GLNPO-GLISP yield highly comparable results for major ions and trophic state parameters (nutrients, chlorophyll a) in Lake Michigan. Statistically significant differences in mean values were found for a few parameters, but the absolute differences in the mean values were very small (e.g., for nitrate, 0.274 ± 0.02 mg N/L [EMAP] versus 0.287 ± 0.011 mg N/L [GLNPO-GLISP]). That the two programs yield essentially the same results indicates that laboratory accuracy and precision are very good for both programs, but aside from this useful operational finding, the results are not surprising.

The programs had comparable numbers of sampling stations (12 for EMAP, 11 for GLNPO-GLISP) and both achieved broad coverage of the lake's offshore region. The EMAP stations were uniformly spaced on a grid established by a randomly selected starting point. According to EPA (1994b), the GLNPO-GLISP sites were not selected on a probabilistic basis, but the exact basis for their selection is unclear. However, it is obvious from Figure 2b

Resource Components 83

in EPA (1994b) that broad, relatively uniform coverage of the offshore region was sought. It is unlikely that the exact locations of the GLNPO-GLISP sites were based on specific information about the sites; to this extent the selection probably was random (within the constraint of adequate geographic coverage). Thus, there appears to be little need to restrict sampling to the arbitrarily selected EMAP grid points within a large, relatively homogeneous water body like Lake Michigan. EMAP should be able to use the historical database and the existing sampling program of EPA's Great Lakes National Program Office (GLNPO) under the Great Lakes International Surveillance Plan (GLISP) to assess offshore trophic conditions in Lake Michigan.

Several similar types of comparisons were made between trophic-state data collected by EMAP on Lake Superior in spring of 1992 and various data collected by Environment Canada. Interpretation of the comparisons was complicated by temporal differences in sample collection and geographic distribution of sampling sites. In general, however, differences between EMAP results and other data sets were quite small, even when differences in mean values were statistically significant.

A limited analysis of the appropriate index period for sampling trophic status was described using Environment Canada's data for spring and fall 1992 in Lake Superior and past data from GLNPO-GLISP for Lake Michigan. It is unclear why EMAP did not collect trophic-state data seasonally on Lake Michigan in 1992 during the summer and fall cruises mentioned in the status report. The report concludes that spring is a better index period in which to measure trophic state because nutrient levels decline in summer (as a result of primary production). Although this is undoubtedly true, few limnologists would agree that trophic state should be measured based on one sampling time per year. The year-to-year variance in chlorophyll *a*, as well as in nutrient levels, probably is higher in a spring index period because of interannual differences in the onset of spring warming and suitable growth conditions for algae. Moreover, if sampling is conducted early in spring, chlorophyll *a* levels may be quite low and unrepresentative of conditions during the maximum growth period for algae.

A considerable effort was made in the pilot study report at attempting to define a depth contour to separate nearshore and offshore sites. According to the authors, values of 85 m and 149 m have been used as the delineating depths in Lakes Michigan and Superior, respectively. Water-chemistry data were not conclusive in confirming this contour for either lake. The authors also used past data on the depth distribution of the benthic invertebrate *Diporeia* to evaluate the nearshore and offshore depth cutoff. *Diporeia* is an amphipod crustacean that is the dominant benthic species in Lake Michigan at depths greater than 30 m. The authors conclude that the *Diporeia* data do not contradict the selection of an 85-m contour for Lake Michigan, but there is no support for this conclusion in the reported data (cf. figures 8 and 9 of EPA, 1994b).

Studies also were undertaken in 1992 to evaluate sampling gear and the index period for sampling benthic invertebrates. A Ponar dredge was found to be more reliable than a box corer, especially for sediment with a high sand content. Results for total abundance of benthic invertebrates, abundance of *Diporeia*, and species richness were found to be similar during the three seasons (spring, summer, fall), and for logistical reasons, EMAP selected summer as the index period for benthic sampling.

Initial work was undertaken on the development of a benthic index using statistical analyses of historical data, but no conclusive results were presented. The report also described EMAP's thinking with regard to diatom-based indicators for biological integrity and trophic condition. EMAP proposes to use a paleolimnological approach in which diatom stratigraphy in long sediment cores will be used to infer background conditions for various lake regions. Present and future conditions will be determined by analyzing diatoms in surface sediment samples or in seston collected in seston traps (a decision on which sampling approach to use has not been made), and results will be compared with the indices derived from the sediment studies. This is an interesting and innovative approach for a monitoring and assessment program, but this somewhat complicated and indirect approach may not be necessary, given the long and rich history of diatom studies on the Great Lakes.

Conclusions

The Great Lakes are by far the largest reservoir of fresh water in North America and collectively constitute about 20 percent of the total global reservoir of fresh water. The lakes are important as a water supply and recreational resource for over tens of millions of people in the eastern United States and Canada, and their economic importance is difficult to overestimate. Nonetheless, it is not clear that EMAP should invest a significant portion of its limited financial and human resources to developing its own program on the Great Lakes. The basis for this conclusion in part lies in the fact that substantial monitoring efforts are already under way, and have been for several decades, by U.S. and Canadian agencies on open-water regions of the lakes, and many state and local agencies conduct monitoring programs in nearshore areas and various harbors and bays. Although current monitoring programs may be inadequate and may need better coordination, it seems that a more efficient approach would be to build on existing monitoring efforts, especially those of EPA's GLNPO, NOAA's Great Lakes Environmental Research Laboratory, the National Biological Survey, and Environment Canada's programs, than to create yet another program. With only a modest investment of funds and personnel, EMAP could serve an important role in stimulating these existing programs, with their inhouse expertise and physical facilities, to expand and coordinate their monitoring and assessment efforts.

Based on the draft strategy document and the 1992 Great Lakes status report, it is not clear that EMAP's vision for a long-term monitoring and assessment program on the Great Lakes is sufficiently forward-looking and comprehensive to address the interests and concerns of policy and decision makers and managers of the Great Lakes ecosystems. The two assessment end points EMAP has selected for the Great Lakes, biotic integrity and trophic condition, encompass only a subset of the major issues of concern about these lakes.

Finally, the grid-based approach that EMAP used in its 1992 pilot studies on Lakes Michigan and Superior is not an efficient or appropriate approach for sampling in large bodies of water like the Great Lakes. For the relatively homogeneous open-water region,

EMAP demonstrated that the existing network of sampling stations is adequate to characterize the resource, and it makes sense to continue long-term monitoring at these sites rather than start a database at new sites based on an arbitrary grid. EMAP did not conduct a comprehensive water-column sampling in nearshore regions in its 1992 field studies, but based on prior information, it concluded that higher spatial heterogeneity in this region would require an enhancement of the grid (increased density of sampling sites) to adequately characterize the resource (e.g., estimate the portion of nearshore waters with acceptable trophic state). However, the spatial variability of water quality is not random in the Great Lakes or in any other large water body. Just as important, a large volume of previous studies, as well as basic land-use, demographic, and geographic data, is available to direct sampling efforts.

For example, if EMAP wishes to know the extent to which the Milwaukee urban area influences nutrient concentrations and related biological response variables in the nearshore region of western Lake Michigan, it would be most inefficient to approach the problem as though it knew nothing about the direction of the sources of influence and therefore set up a random sampling program throughout the region. Instead, it makes sense to recognize that the nutrients are coming from a rather narrow zone along the western shore where the Milwaukee urban area is located. Any reasonable aquatic scientist would approach this problem by gathering information on the sources of influence (major rivers, streams, sewage, and stormwater outfalls) and then setting up a sampling scheme that used this information. Most likely, that would involve sampling in linear transects away from the source area(s).

EMAP states that it is not interested in characterizing local problems, and understandably so. Thus it may not wish to develop an intense enough monitoring network or set of transects to characterize the zone of Milwaukee's influence with an accuracy that would answer all the questions that local and state water-quality managers and policy makers may have. Nonetheless, the issue is relevant to EMAP's stated interests regarding the Great Lakes. EMAP wishes to characterize the nearshore region with known confidence. Conditions in that region are not distributed randomly but with known spatial biases. It makes sense to use available information and design a monitoring program accordingly.

Probabilistic sampling has a role in EMAP—perhaps even in a Great Lakes monitoring program. However, complete reliance on random site selection is unnecessary and inefficient. In the above example, a random component could enter the site selection process in terms of a list frame of the onshore sources (rivers, sewage outfalls, cities) above a certain magnitude that are likely to affect the nearshore region. Financial and other limitations will not allow detailed transects to be sampled on all possible source areas, and selection of those to be monitored would be done randomly.

SURFACE WATERS
(Executive Summary, from NRC, 1994b)

This third report reviews the EMAP-Surface Waters monitoring component in the context of the larger program. This report pays particular attention to the strengths and weaknesses of the overall program, as they affect EMAP-Surface Waters. These program-wide issues fall into three major classes: assessment end points; indicators; and design. This report includes a review of the Lakes Pilot Project and on early information available on the streams program.

The EMAP-Surface Waters group should be commended for its investigation into the critical ways different sources of variation will affect EMAP's ability to detect status and trends. The EMAP-Surface Waters Implementation pilot was reasonably organized, and logistical aspects of the operation were well planned. Execution of the field portion of the regional assessment of the pilot was successful and valuable experience was gained in the site selection process and in evaluating the logistical aspects of the program.

Although the pilot project was wisely question-driven, some of the questions are unclear or inadequate. In general, the pilot study could be substantially improved, not just because it failed to address some of the questions and goals it set for itself, but also because those goals and questions are a very incomplete list of the fundamental issues that need to be addressed before the surface waters program is ready for full implementation. In particular, issues of coordination among resource groups, relationships between indicators and specific stressors, and ability to make inferences on scales ranging from single lakes to entire regions were not addressed. Not every issue can be addressed in a single pilot study, but there appears to be no overall plan to address these issues sequentially in subsequent pilot studies. In addition, oversight and involvement of senior scientists from a central management team at EMAP Center might have enhanced the scientific rigor of the pilot study, improved the design, analysis and reporting phases of the pilot study, and produced more useful models for the full program.

Background and Objectives

The Surface Waters component of EMAP has responsibility for achieving EMAP goals for the nation's lakes, reservoirs, streams, and rivers. Surface Waters is one of eight EMAP resource groups. The other resource groups are: forests, estuaries, agro-ecosystems, arid lands, the Great Lakes, wetlands, and landscape ecology. EMAP-Surface Waters' initial efforts emphasized lakes and reservoirs, and this portion of the program is more fully developed than the program for rivers and streams. For lakes and reservoirs a pilot project was conducted from 1991 to 1993 in the northeast area of the United States. A stream pilot project was conducted in the mid-Appalachian area in 1993.

EMAP-Surface Waters differs from most other surface-water monitoring approaches in that it is statistically designed to infer information from a sample of lakes to the entire population of lakes on regional and national scales.

Objectives of EMAP-Surface Waters parallel those of the general EMAP program. The objectives of the Surface Waters Component (from D. McKenzie, EMAP Program Officer, verbal communication, February 24, 1994) are to:

Resource Components

- estimate the current status, trends, and changes in selected indicators of condition of the nation's lakes, reservoirs, streams, and rivers on a regional basis with known confidence;
- estimate the extent (number and surface area of lakes and reservoirs, miles of rivers and streams) of the nation's lakes, reservoirs, streams and rivers with known confidence;
- seek associations between selected indicators of natural and anthropogenic stresses and indicators of the condition of ecological resources; and
- provide annual statistical summaries and periodic assessments on the condition of the nation's lakes, reservoirs, streams and rivers.

Assessment End Points

EMAP-Surface Waters has designated three assessment end points for the lakes portion of their program: biological integrity; trophic condition; fishability.

The choice of assessment end points provides the foundation for the EMAP Lakes Program. This first step, therefore, is of critical importance. The EMAP-Surface Waters' current selection of end points needs further definition and improvement. Of the three end points, biological integrity is the most problematic. As used by EMAP-Surface Waters, this term is vague and all-inclusive, conceptually subsuming the content of the other end points and all other more specific environmental problems in lakes. Such a broadly defined term may be useful in summarizing diverse data or in addressing the multiple issues related to environmental quality, but it is not specific enough to be a useful end point upon which to design data monitoring activities. Therefore, EMAP-Surface Waters (and other EMAP resource groups) should use the term "appropriate biological diversity" instead of "biological integrity" as an assessment end point, as discussed in Chapter 2. This term is based on objective evaluations and depends on measurable, quantifiable reference systems, and its use should lead to the development of objective, quantifiable indicators.

The other two assessment end points for EMAP-Surface Waters are trophic condition and fishability. In theory, each could be defined in reasonably unambiguous ways and straightforward means can be developed to measure them quantitatively. Nonetheless, further efforts are needed to refine the definitions and measurement strategies for both end points.

In addition, although EMAP's financial resources will be limited, it is imprudent to exclude drinking water from consideration as a societal value in its surface water assessment program. The EPA and the revised Clean Water Act both express and affirm the concept of holistic watershed planning and management. Also many impoundments and natural lakes are used both for recreation and for drinking water supplies. This is an example where close cooperation with EPA's Office of Water 305b program could be very beneficial.

Indicators

Once the major assessment end points have been decided, the next critical task is to determine what measurements are necessary to assess these end points. When the problem has been stated, a conceptual model of how the particular system works with respect to the problem should then be stated explicitly. Examination of the conceptual model leads to the selection of potential indicators, which are tested in the field. The indicators are selected on the basis of known or suspected cause-effect relationships that are identified in the conceptual model. Until March 1994, EMAP provided no satisfactory *program-wide* guidelines for indicator selection strategy, and each resource group was left to fend for itself with little or no guidance from EMAP-Center. As a result, use of conceptual models to drive indicator selection is not well developed in EMAP.

The conceptual model implicit in the EMAP-Surface Waters strategy document underestimates the complexity of freshwater ecosystems. There is no consideration of factors like biogeography, seasonal shifts in community structure with secondary nutrient depletion, competition, predation, or hydrologic factors.

Resource Components

Therefore, explicit conceptual models of the ecological systems being monitored should be used to guide indicator development.

EPA's peer-review panel was concerned about heavy reliance on indices with unknown properties. Use of indices to describe complex ecosystems has some advantages but also some important disadvantages. The major advantage is the ability of an index to condense many parameters into a single number, which at first glance may be easier to understand. A major disadvantage is that the statistical properties of the index are often not well understood; moreover the indices are often nonlinear; that is, a change from 1 to 2 is not the same as a change from 2 to 3.

Rather than relying upon a univariate index with unknown statistical properties, it is possible to use the multi-response vector of the original parameters and apply multivariate statistical techniques for analysis (e.g., logistic regression, clustering and pattern-recognition algorithms, neural network analysis), or exploratory data techniques involving better visualization of multi-dimensional data (Becker et al. 1987, Cleveland and McGill 1988). Nonparametric multivariate procedures also exist (as in Zimmerman et al. 1985) for testing whether groups of multivariate data points are significantly different from each other (e.g. comparing disturbed to undisturbed areas). EMAP-Surface Waters should continue its efforts to develop indices using a number of different approaches including multivariate statistical and exploratory data analyses. In addition, appropriate new statistical methods should be incorporated as they become available.

Sampling Design

The design for the Surface Waters component follows the overall EMAP design. The scheme for lakes is better developed than that for streams, which have not received detailed consideration to date. About 3,200 lakes will be selected using a probability-based sampling scheme. A different subset of 800 of them will be sampled each year so that every lake is sampled once every four years. Lakes will be stratified into size classes so that large lakes (which are relatively rare compared to small lakes) are represented in the sample. The random selection of the lakes will occur in a way that maintains uniform spatial coverage nationally. There are several areas of concern regarding the general EMAP sampling design.

A watershed perspective is lacking in the sampling design. Because surface water systems are linked with their watersheds, the lack of a watershed perspective will severely limit the identification of likely causes of detected changes in the EMAP lakes. Without this watershed perspective, landscape characterization data cannot be used to evaluate the status of individual aquatic resource units. Thus, a greater emphasis should be given to concomitant measures of watershed characteristics. Remotely sensed data on land use and cover could be used to great advantage. Representatives of EMAP have recently indicated (May 1994, conference call with surface water panel members) that they will be using a watershed approach for their data gathering, and the committee encourages this approach.

Another concern is that the sampling design may not be sensitive enough to detect a change in condition unless the change is very large in magnitude and affects most lakes and streams in a region. There are many types of lakes and streams in many types of landscapes and each one has different sensitivities to a particular stress. It is not clear that enough sensitive lakes and streams will be included in the sample to detect a change due to a particular stress. Because lakes and streams will be sampled during one 9-week period, some measurements may not be made at the biologically most meaningful time, thus decreasing the sensitivity of the measurement.

In addition, the sampling design may have difficulty detecting changes in biological measurements over time. Variances of biological populations (and therefore community measurements) among lakes and within lakes over the course of a year are large.

EMAP-Surface Waters has been a leader in performing tests of statistical power to detect changes or differences with real limnological data collected by various state agencies. Power studies to date have examined primarily physical and chemical variables since these data were relatively available. Similar tests with published or simulated biological (population and community level) data should be vigorously pursued, because EPA indicates that these data will be important in formulating indicators similar to Karr's Index of Biological Integrity (Karr, 1992).

It is not clear how useful the trends that EMAP may detect will be, and whether EMAP will be able to relate such trends to specific stressors is uncertain. Because of the four-year revisitation rate, the current design essentially does not allow for site-specific inferences to be made. Although it is not an explicit goal of EMAP to make site-specific inferences, there is value in making site-specific inferences from well-chosen sites. This would augment the basic EMAP design and the added value could be achieved at small additional cost. Therefore, a substantial number of sites should be sampled annually. Some of these sites might be selected because they are known or suspected to be sensitive indicator lakes for selected stressors.

Additional power tests should be performed to examine the ability of the current design to detect status and trends for quantiles in the tails of distributions (e.g., lower and upper 10th percentiles).

Lake Pilot Project

The surface water component of EMAP began its first year pilot study during the summer of 1991. Pilot activities included a regional sampling effort (EPA Region 1), a set of more focused indicator-development studies, and an analysis of the effects of different types and magnitudes of variability on the ability to detect regional trends.

The regional assessment portion of the pilot study represents the first application of the general EMAP design to surface water ecosystems. The EMAP-Surface Waters implementation pilot was reasonably organized, and logistical aspects of the operation were well planned.

Field execution of the regional assessment portion of the pilot was successful. Valuable experience was gained in the site selection process and in evaluating the logistical aspects of the program. However, a substantial portion of the data was not analyzed in time to meet deadlines for the pilot study report. This suggests that a larger investment in data analysis will be necessary if a larger scale implementation is to be completed in a timely fashion.

The design of the indicator development study was not as good as that of the regional assessment portion of the pilot. The scope of this portion of the pilot was too ambitious given the financial resources available. The response of lakes to catchment disturbance, or even the ability of certain indicators to detect the response, is unlikely to be discerned without a much larger set of lakes selected specifically to address this question. With only four to six lakes per class it was unreasonable to expect to be able to see a strong signal between disturbance and the response variables.

Field sampling for the indicator portion of the pilot appeared to go smoothly. Useful variance estimates and time and cost estimates were obtained for many of the assemblage indicators. However, there appear to be major difficulties with the analyses of the indicator assemblage data. They include:

- lack of planning and coordination;
- lack of statistically sound hypothesis evaluation; and
- lack of any quantitative comparison between the various indicator variables measured.

Streams

EMAP-Surface Waters also began to conduct a pilot program on streams in the summer of 1993. It is difficult to evaluate this pilot study, because of the scarcity of documentation.

The sampling strategy to be used in the stream survey needs further development. Based on the limited information now available, it was premature to embark on a stream pilot study at this time. The currently conceived sampling strategy appears inadequate to characterize stream quality either chemically or biologically. Not everything can be planned in advance, and there still is room for the trial-and-error approach in developing large-scale programs like EMAP. Nonetheless, the scale of financial and human resources required even for a pilot-level survey is sufficiently great that EPA must not only minimize the risk of error, but also maximize the likelihood that it will successfully address the critical issues necessary for planning a full-scale stream survey. EMAP is not presently in this position.

EMAP should decide what its overall objectives are for assessing the status of the nation's rivers and streams. These objectives—and a strategy to achieve them—need to be developed within the context of existing federal monitoring programs. This should occur before EMAP proceeds with the development of stream pilot studies. The currently conceived sampling strategy is not appropriate to characterize stream quality either chemically or biologically.

It is unclear to what extent there has been substantive involvement of the scientific community in the planning done to date for the streams pilot. Therefore, EMAP-Surface Waters' scientists should spend time developing a substantive planning document and continue their dialogue with stream scientists in other branches of EPA, in other water-related federal agencies (U.S. Geological Survey, U.S. Fish and Wildlife Service, etc.) and in the academic community to better evaluate how the stream phase of EMAP should be designed.

Intra-Agency Cooperation

Much routine water quality sampling done by state pollution control agencies on surface waters nationwide is funded through EPA's Office of Water under Section 305b of the Clean Water Act. Closer collaboration between the 305b program and EMAP has the potential to enhance the effectiveness of both programs while reducing the overall cost of federal monitoring programs for surface water quality.

Therefore, EMAP-Surface Waters and the EPA Office of Water should work together to insure that data collected under the 305b program can be useful not just for compliance monitoring (the primary focus of current programs in most states), but also to assess temporal and geographic trends in water quality.

Oversight And Coordination Among EMAP Resource Groups

Coordination among resource groups is especially important for the Surface Waters component of EMAP. Surface waters are affected by processes occurring within the terrestrial ecosystems in their watersheds. Currently, the Surface Waters group is analyzing riparian vegetation. However, it is not clear that the classification system being used is the same as that used by terrestrially-focused resource groups. Without a closer interaction with the terrestrial components of EMAP, an opportunity for comprehensive understanding of how and why lakes may be changing is likely to be missed.

EMAP-Center Organization

There is a continuing lack of a clearly defined procedure for defining and prioritizing the assessment questions that can and will be addressed by the program. These questions are critically important, because they will drive the sampling strategies and clarify the goals of EMAP-Surface Waters and the other resource groups.

A procedure should be developed to identify the most important assessment questions from a policy perspective, but at the same time ensure that it is scientifically feasible to address the questions. One possible approach is to formalize a planning structure that would be composed of guidance panels associated with each resource group. A central planning committee, composed of representatives from each of the thematic panels would then make the hard decisions about resource allocation and attempt to optimize efficiency and coordination between groups. Some aspects of such a planning structure already exist within EMAP. However, it is critical that the guidance panels also include representatives of EMAP clients, i.e., policy makers and the larger scientific community. Most panel members should be external to EMAP and to EPA and they should be leaders in their areas of expertise. These panels should not duplicate the advisory and planning functions of the current peer-review panels or of EPA's Science Advisory Board.

References

American Statistical Association (ASA) Committee on EMAP. 1992. Review of EMAP Statistics and Design. Report of a meeting November 4-6, 1991, San Francisco, CA. American Statistical Association, Alexandria, VA.

Barber, M. C., Ed. 1994. Environmental Monitoring and Assessment Program Indicator Development Strategy. EPA/620/R-94/XXX. EMAP Center, U.S. Environmental Protection Agency, Research Triangle Park, N.C.

Becker, R. A., W. S. Cleveland, and A. R. Wilks. 1987. Dynamic graphics for data analysis. Statistical Science 2(4):355-395. See also five "Comment" articles plus rejoinder in the same issue.

C. L. Campbell, J. Bay, C. D. Franks, A. S. Hellkamp, G. R. Hess, M. J. Munster, D. A. Neher, G. L. Olson, S. L. Peck, J. O. Rawlings, B. Schumacher, and M. B. Tooley. 1993. Environmental Monitoring and Assessment Program Agroecosystem Pilot Field Program - 1993. U.S. Environmental Protection Agency, Office of Research and Development, Washington, D.C.

C. L. Campbell, J. M. Bay, A. S. Hellkamp, G. R. Hess, M. J. Munster, K. E. Nauman, D. A. Neher, G. L. Olson, S. L. Peck, B. A. Schumacher, K. Sidik, M. B. Tooley, and D. W. Turner. 1994. Environmental Monitoring and Assessment Program Agroecosystem Pilot Field Program Report - 1992. U.S.

Environmental Protection Agency, Office of Research and Development, Washington, D.C.

Cleveland, W. S., and M. E. McGill. 1988. Dynamic Graphics for Statistics. Wadsworth and Brooks/Cole, Belmont, California.

EPA. 1991. An Overview of the Environmental Monitoring and Assessment Program. EMAP Monitor, January 1991. EPA-600/M-90/022. U.S. EPA, Office of Research and Development, Washington, D.C.

EPA. 1992a. Environmental Monitoring and Assessment Program Great Lakes Monitoring and Research Strategy. Environmental Research Laboratory, Office of Research and Development. U.S. EPA, Duluth, MN.

EPA. 1992b. Final Peer Review Panel Report of Environmental Monitoring and Assessment Program - Great Lakes Monitoring and Research Strategy. Office of Environmental Processes and Effects Research, U.S. EPA, Washington, D.C.

EPA. 1992c. Environmental Monitoring and Assessment Program Great Lakes Monitoring and Research Strategy Response to Peer Review Panel Report. Environmental Research Laboratory, Office of Research and Development. U.S. EPA, Duluth, MN.

EPA. 1993. Environmental Monitoring and Assessment Program Guide, EPA/600/XX-93/XXX. U.S. EPA, Atmospheric Research and Exposure Assessment Laboratory, Research Triangle Park, NC.

EPA. 1994a. Landscape Monitoring and Assessment Research Plan, EPA 620/R-94-009. U.S. EPA, Office of Research and Development, Washington, D.C.

EPA. 1994b. Environmental Monitoring and Assessment Program for the Great Lakes. FY92 Status Report. Environmental Research Laboratory, Office of Research and Development. U.S. EPA, Duluth, MN.

EPA Science Advisory Board. 1988. Future Risk: Research Strategies of the 1990s. SAB-EC-88-040. Science Advisory Board, U.S. EPA, Washington, D.C.

Foley, G. J. 1994. Letter and attachment on EMAP response to NRC Forests and Estuaries report (see NRC, 1994a).

Fore, L., J. Karr, and L. Conquest. 1994. Statistical properties of an index of biological integrity used to evaluate water resources. Canadian Journal of Fisheries and Aquatic Sciences 51:1077-1087.

Heck, W. W., C. L. Campbell, R. P. Breckenridge, G. E. Byers, C. M. Hayes, G. R. Hess, V. M Lesser, J. R. Meyer, T. J. Moser, D. A. Neher, G. L. Olson, S. L. Peck, J. O. Rawlings, and C. N. Smith. 1991a. Environmental Monitoring and Assessment Program Agroecosystem Research Plan. Peer Review Draft. February 1, 1991. (The document contains no publication information.)

Heck, W. W., C. L. Campbell, R. P. Breckenridge, G. E. Byers, A. L. Finkner, G. R. Hess, J. R. Meyer, T. J. Moser, S. L. Peck, J. O. Rawlings, and C. N. Smith. 1991b. Agroecosystem Monitoring and Research Strategy. EPA 600/4-91/013. U.S. Environmental Protection Agency, Office of Research and Development, Las Vegas, Nevada.

Heck, W. W., C. L. Campbell, A. L. Finkner, C. M. Hayes, G. R. Hess, J. R. Meyer, M. J. Munster, D. A. Neher, S. L. Peck, J. O. Rawlings, C. N. Smith, and M. B. Tooley. 1992. Environmental Monitoring and Assessment Program Agroecosystem 1992 Pilot Project Plan. U.S. Environmental Protection Agency, Office of Research and Development, Washington, D.C.

Houghton, R. A., J. E. Hobbie, J. M. Melillo, B. Moore, G. J. Peterson, G. R. Shaver, and G. M. Woodwell. 1983. Changes in the carbon content of terrestrial biota and soil between 1860 and 1980: a net release of CO_2 to the atmosphere. Ecological Monographs 53:235:262.

Hunsaker, C. T., and D. E. Carpenter. Eds. 1990. Ecological Indicators for the Environmental Monitoring and Assessment Program. EPA 600390060. U.S. Environmental Protection Agency, Office of Research and Development, Research Triangle Park, N.C.

Karr, J. R. 1992. Measuring biological integrity: lessons from streams. Pp. 83-104 in Ecological Integrity and the Management of Ecosystems, S. Woodley, J. Kay, and G. Francis eds. St. Lucie Press, Delray Beach, Fla.

Larsen, H. P., N. S. Urquhart, and D. L. Kugler. 1993. Regional scale monitoring of indicators of trophic condition of lakes. Water Research.

Lesser, V. M., and W. S. Overton. 1994. EMAP Status Estimation: Statistical Procedures and Algorithms. EPA 620/R-94-008.

McDonnell, M. J., and S. T. A. Pickett, eds. 1993. Humans as Components of Ecosystems. Springer-Verlag, New York. 364 p.

Memorandum of Understanding between the U.S. Environmental Protection Agency Office of Research and Development and National biological Survey. September 30, 1994.

Messer, J. J., R. A. Linthurst, and W. S. Overton. 1991. An EPA Program for Monitoring Ecological Status and Trends. Environmental Monitoring and Assessment 17:67-78.

NRC. 1992. Review of EPA's Environmental Monitoring and Assessment Program Interim Report. National Academy Press, Washington, D.C.

NRC. 1993a. Research to Protect, Restore, and Manage the Environment. National Academy Press, Washington, D.C.

NRC. 1993b. Issues in Risk Assessment. National Academy Press, Washington, D.C.

NRC. 1994a. Review of EPA's Environmental Monitoring and Assessment Program: Forests and Estuaries. National Academy Press, Washington, D.C.

NRC. 1994b. Review of EPA's Environmental Monitoring and Assessment Program: Surface Waters. National Academy Press, Washington, D.C.

Oliver, C. D., and B. C. Larson. 1990. Forest Stand Dynamics. New York: McGraw Hill.

Olsen, A. R. Ed. 1992. The Indicator Development Strategy for the Environmental Monitoring and Assessment Program. EPA

References

600391023. U.S. Environmental Protection Agency, Environmental Research Laboratory, Corvallis, Oregon.

RAF (Risk Assessment Forum). 1992. Framework for Ecological Risk Assessment. EPA 630R92001. U.S. Environmental Protection Agency, Risk Assessment Forum, Washington D.C.

Saila, S. 1993. The use of multivariate trend analysis to provide preliminary, multispecies management advice. Pp. 493-506 in G. Kruse, D. M. Eggers, R. J. Marasco, C. Pautzke, and T. J. Quinn II, eds. Management Strategies for Exploited Fish Populations. Alaska Sea Grant College Program report AK-SG-93-02, Alaska Sea Grant, Fairbanks, Alaska.

Stanley, D. W. 1993. Long-term trends in Pamlico river estuary nutrients, chlorophyll, dissolved oxygen, and watershed nutrient production. Water Resources Research 29:2651-2662.

Stehman, S. V., and W. S. Overton. 1994. Comparison of variance estimators of the Horvitz-Thompson estimator for randomized variable probability systematic sampling. J. Amer. Statistical Ass. 89:30-43.

Stevens, D. L. 1993. Implementation of a national monitoring program. Journal of Environmental Management 42:1:1.

Thornton, K. W., G. E. Saul, and D. E. Hyatt. 1994. Environmental Monitoring and Assessment Program Assessment Framework. EMAP Center. U.S. Environmental Protection Agency, Research Triangle Park, N.C.

Turner, B. L. II. (ed). 1990. The Earth as Transformed by Human Action. Cambridge University Press with Clark University. Cambridge, Massachusetts. 713 p.

Urquhart, N. S., W. S. Overton, and D. S. Birkes. 1993. Comparing Sampling Designs for Monitoring Ecological Status and Trends: Impact of Temporal Patterns. Pp. 71-86 in V. Barnett and K. F. Turkman, eds. Statistics for the Environment. John Wiley & Sons, Ltd., London.

Weisberg, S. B., J. B. Frithsen, A. F. Holland, J. F. Paul, K. J. Scott, J. K. Summers, H. T. Wilson, R. Valente, D. G. Heimbuch, J. Gerritsen. S. C. Schimmel, and R. W. Latimer.

1992. EMAP-Estuaries Virginian Province 1990 Demonstration Project Report. EPA 600/R-92/100. Environmental Research Laboratory, U.S. EPA, Narragansett, Rhode Island.

Zimmerman, G. M., H. Goetz, and P. W. Mielke, Jr. 1985. Use of an improved statistical method for group comparisons to study effects of prairie fire. Ecology 66:606-611.

Appendix A

September 20, 1994 letter from
Dr. Edward Martinko, Director, EMAP

Appendix A

UNITED STATES ENVIRONMENTAL PROTECTION AGENCY
WASHINGTON, D.C. 20460

SEP 20 1994

OFFICE OF
RESEARCH AND DEVELOPMENT

Dr. Richard Fisher
c/o The Committee to Review the EPA's
  Environmental Monitoring and Assessment Program
Water Sciences and Technology Board
National Research Council
2101 Constitution Avenue, NW
Washington, DC 20418

Dear Dick:

    I am pleased to provide you with an overview of important changes made in the operations and management of the Environmental Monitoring and Assessment Program primarily as a result of deliberations by the National Research Council's review, and to a lesser extent other reviews of the Program or its various components. The overview lays out in an organized manner responses to concerns voiced by reviewers, concerns that have helped to focus our attention in strengthening the scientific and administrative bases of our long-term, national endeavor.

    I look forward to seeing recommendations for further improvement of our activities.

Sincerely,

Edward Martinko, Ph.D.
Director
Environmental Monitoring and
Assessment Program [8205]

cc: Robert J. Huggett, AA/R&D [8101]
      H. Hatthew Bills, OMMSQA [8201]
      Rick Linthurst, EMAP-Center
      Jay Messer, AREAL-RTP
      Sidney Draggan, EMAP-HQ [8205]

Appendix A

Précis

As you know, the Environmental Monitoring and Assessment Program's (EMAP) scientific and managerial foundations have evolved dramatically and demonstrably over the course of the past five years. These important changes are in response to research findings within the program and to the National Research Council's (NRC) review. This evolution has caused some misunderstandings about what has and what has not been altered. As you finalize your review of the EMAP, the NRC Committee staff and I believed that it would be useful to summarize the changes made in the Program attributable to the written reviews and discussions with the NRC Committee.

I have organized into categories what I believe are the important changes that represent areas of expressed NRC interest. Within each category, I have attempted to describe what has changed or what is changing. I believe our actions are consistent with the specific comments or recommendations of the Committee. As the categories are not completely independent, there is some duplication in the statement of actions and changes.

I hope you will find this summary useful in preparing the final report on EMAP. The overview categories include:

- Indicator Development;
- Sampling Design;
- Trend Detection;
- Landscapes;
- Analysis of EMAP Data, particularly Estuaries, and Publication of Results;
- Information Management;
- Assessment, Integration and Coordination;
- Cause and Effect Relationships;
- Program Management;
- Involvement of Scientific Community and Peer Reviews; and
- Inter- and Intra-Agency Cooperation.

Selected EMAP Changes Occurring From 1992 To The Present

Indicator Development

EMAP instituted an investigator-initiated indicator development research grants program. This grants program is independent of the activities of the EMAP's individual Resource Groups. The amount invested was $3.0M in Fiscal Year (FY) 1994 and $4.0M is anticipated in FY 1995.

Currently, we have revised and completed the Program's indicator development strategy to ensure a consistent approach to indicator identification, testing and validation, and incorporation into the monitoring program. The need for conceptual models to drive the indicator selection process is specifically emphasized and all Resource Groups are re-evaluating their models for a review within the next twelve months.

A permanent Indicator Development Coordinator has been appointed this year.

Also, we have selected and defined common environmental values for all Resource Groups to promote program integration. There is still considerable work to do, however, on the indicators and measurements necessary to quantify those values. This work will continue over the next few years before significant progress in quantification capability is expected.

The Committee expressed concern over the variability of the indicators within EMAP-Estuaries, Surface Waters and Forests. In response, each of these Resource Groups is now closely evaluating the variability associated with all indicators collected to date in each of their regional studies. After completion, these results will be examined closely by external, independent scientists to determine if the approach is viable. These reviews will be conducted through the mail primarily with panel reviews scheduled as deemed necessary.

EMAP-Surface Waters is supporting research on both lake and stream indicators through eight cooperative agreements with universities and under two agreements with Federal agencies.

Specifically for EMAP-Estuaries, the following changes have been made or initiated:

- EMAP-Estuaries has developed, and will be publishing, an explicit conceptual model to enhance direction of its indicator research program; to improve capabilities to explain its approach; and, to provide integration across multiple estuarine Provinces.

- EMAP-Estuaries is re-examining fish trawl, tissue contaminant, and pathology surveys as viable and reliable indicators. We are pursuing a joint research program with the National Oceanic and Atmospheric Administration (NOAA) to re-evaluate these indicators and field protocols in 1995-1996. All routine monitoring has ended in 1994 so that these resources could be invested in data analyses and for further indicator development. Indicator development and testing in the estuarine systems will be the primary effort for 1995 and beyond;

Appendix A

- This Resource Group is working through Interagency Agreements with the National Biological Survey (NBS) and the National Wetlands Inventory to examine the utility of various remotely-sensed indicators with regard to wetlands, submerged aquatic vegetation beds, and offshore (that is, nearshore ocean) chlorophyll concentrations. The group has developed a Cooperative Agreement with the Marine Resource Institute in St. Petersburg, Florida to develop indicators of ecosystem function with special emphasis on productivity and eutrophication. Results will be forthcoming, however, within approximately two years;

- The group is examining the validity of the Benthic Index from multiple perspectives: geographic applicability to the Northeast, West Coast, and South Florida; the use of weighting factors in the construction of the Index and the sensitivity of the index to these weightings; and, the construction of National paradigms using benthic indicators (in addition to the regional indexes now developed). Work on the East and Gulf Coasts will be complete by 1996. These results will be used to determine the utility, stability and reliability of a Benthic Index;

- The group is conducting research at the Gulf Breeze Laboratory and under an Interagency Agreement with NBS to evaluate and separate the potential causes of observed fish pathology. The Interagency Agreement examines the response of macrophage aggregates to environmental stresses (for example, contaminants and hypoxia) through rearing studies and investigations of the role of macrophage aggregates in the development of neoplasia. Results will be used to assist in the selection of additional indicators of fish stress in the field;

- EMAP-Estuaries supports investigator-initiated research through portions of nine Cooperative Agreements with the University of Rhode Island at Narragansett, Rhode Island; the University of North Carolina at Wilmington in Wilmington, North Carolina; the Marine Resources Research Institute in Charleston, South Carolina; the Marine Research Institute in St. Petersburg, Florida (two Cooperative Agreements); the Skidaway Institute for Oceanography in Savannah, Georgia; the Gulf Coast Research Laboratory in Biloxi, Mississippi; and, the University of Mississippi, in Oxford. In addition, the group has supported a student grant for investigator-initiated research at the University of South Carolina's Department of Marine Sciences - Baruch Institute; and

- The group is conducting a detailed analysis of its demonstration projects in the Virginian and Louisianian Provinces to determine alternatives for measuring estuarine condition. These activities involve EMAP personnel, university researchers representative of the areas of investigation, other EPA and NOAA personnel, and State resource agency researchers. All of these estuarine enhancements are expected to facilitate the design of a national estuarine monitoring program, the selection of indicators to meet the objectives of such a program and the data analysis capabilities needed to document with confidence estuarine condition, nationally.

For EMAP-Forests, the following indicator changes have been addressed:

- Currently, EMAP-Forests is early in the process of reviewing the forest ecological condition indicators chosen and tested, to date. Complete analyses are not expected for eight to twelve months. The number of indicators has been increased to fourteen in regional pilots in response to NRC concerns that the indicators chosen are related only to productivity and that more indicators should be chosen and tested carefully;

- The Program will not expand into additional States until existing demonstration projects are evaluated more fully, as recommended by the NRC. These evaluations are currently underway, but will again, take eight to twelve months for completion; and

- Attention is being focused on development of specific conceptual models driving indicator selection, as proposed in the indicator strategy and as recommended.

Sampling Design

EMAP has increased involvement of university statisticians and the monitoring research community in the further development, and evaluation, of the EMAP Statistical Design through ten nationally-competed Cooperative Research Agreements, totaling $0.80M in FY 1994; these will continue in 1995 and 1996.

The EMAP Design Group has developed over the past three years improved implementation strategies and frameworks for sampling estuaries, lakes, streams, and the Great Lakes addressing statistical issues identified by external review committees.

We have conducted comparisons of the statistical efficiency of alternative lake and stream survey sampling designs for estimation of status and trends, verifying that there is more information to be gained by sampling more sites than repeat sampling at a single site.

EMAP policy has been changed to routinely provide an enhanced sample grid to States and to university cooperators who wish to sample more intensively. The States have benefited directly, particularly in EMAP-Estuaries and EMAP-Surface Waters. Each year, we have experienced increasing extramural interest in transferring technology related to reliable and cost-effective sampling protocols.

EMAP-Estuaries and EMAP-Surface Waters have now progressed to the point where they are examining the use of multivariate statistical tools to characterize multiple indicators through multi-dimensional indices. This is an area for future investment in extramural and intramural investigations that will be proposed for funding in 1996.

EMAP Design and Statistics has funded a Cooperative Research Agreement with the University of Wyoming to study the issue of using EMAP data with non-probability-based data. This research will hopefully add a model-based element to the Program to ensure more complete use of existing

Appendix A

data. This design-based research is coming now to a close and after the results are evaluated, the next studies will be designed to address this complex statistical issue.

We have made an investment in university research designed to achieve integration of spatial statistics and Geographic Information Systems for efficient analysis of ecological monitoring information. For example, we are working with Iowa State University in this area; they are funded at $0.91M.

EMAP Design and Statistics has created an ad hoc Geospatial Monitoring Working Group with senior statisticians working at other Federal agencies to explore coordination of surveys by EPA-EMAP, the Soil Conservation Service's National Resource Inventory, the U.S. Department of Agriculture's National Agricultural Statistics Service, the U.S. Forest Service's Forest Inventory and Analysis, the National Biological Survey, and the U.S. Fish and Wildlife Survey's National Wetlands Status and Trends Program. The overall goal of this effort is to seek some common ground for survey designs. Currently, there is no target date for a joint report on progress.

The NRC suggested that some plots in forests—an issue applicable to all the resource groups—should be visited each year. While there is debate by the statisticians about that recommendation, EMAP-Forests has a database with repeat measurements over the last four years at a large number of sites. We are analyzing these data carefully to assess the benefits to be derived from adopting the recommendation and hope to have analyses complete within one year.

EMAP-Forests will work during 1995 to evaluate collected data and to design options for implementing a Forest Monitoring Program.

Detection of Trends

Intensive analysis of EMAP-Estuaries and EMAP-Surface Waters data was initiated in 1994 to address the issue of the ability of EMAP to detect trends, within which time frames, and of which magnitudes. It does appear that most measurements meet the Data Quality Objectives of the Program, although all of the analyses will not be complete for several months.

The Program generated and advertised a Request for Proposals (RFP) to the National Science Foundation's (NSF) network of Long-Term Ecological Research (LTER) and Land-Margin Ecosystems Research (LMER) sites. The objectives of the Request were to stimulate the development and evaluation of: 1) ecological indicators and indices; 2) testable hypotheses relating observed ecological response to natural and anthropogenic stresses; and, 3) models and statistical methods to estimate the status of ecological condition and trends at watershed and regional scales, based on site specific and probability-based sampling data. Research funded through this announcement will begin in FY 1995.

EMAP has funded cooperative studies and grants with intensive site-based sampling programs (see LTER and LMER, above; universities and the U.S. Geological Survey [USGS]) to evaluate the ability to detect trends through individual and combined efforts. Also, these studies are proving useful in terms of identification of indicators and issues relating to extrapolation of site-specific results.

EMAP-Estuaries, EMAP-Surface Waters and the EMAP Design and Statistics Team are funding as well as conducting examinations of the effect of multiple design modifications on the detection of different types of trend functions, including non-linear and step functions. This work will be ongoing and improvements are expected in the near future.

EMAP-Estuaries is examining more fully, as recommended, the use of existing data sets from the Virginian Province to evaluate more completely the utility of indicators, the expected level of trend changes that can be expected over decadal time periods for different indicators, and the spatial intensity needed to characterize individual systems. We are concentrating on data sets from the Chesapeake Bay, Delaware Bay, and Long Island Sound. We will conduct similar analyses with data from the Louisianian Province in 1995-1996 concentrating on data from Apalachicola Bay, Florida; Barataria Bay, Louisiana; and Galveston Bay, Texas.

As with EMAP-Estuaries, EMAP-Surface Waters has made increasing use of existing data sets from State, other Federal agency program, and university research efforts to evaluate intensively the variability of proposed indicators relative to detection of trends. For example, for EMAP-Surface Waters, information from the lake trophic condition databases of Maine, Vermont and New York have been used by EMAP-Surface Waters to evaluate trophic state parameters; and, we have used Dartmouth College's database on zooplankton indicators. All of these studies will be on-going in the foreseeable future.

Landscapes

The original Landscape Characterization Group has been split into two groups. The Landscapes Resource Group is charged with conceptualizing and conducting fundamental as well as applied research on landscape scale indicators (Landscape Ecology) and with validating such indicators for use within the Program. The new Landscape Characterization Coordination Group is tasked with conducting and advancing land cover and land use analyses for use within the Program (and by other interested users).

Recently, this year, we underwent a review of the Landscapes Resource Program by a peer review panel and the Agency's Science Advisory Board. Few adjustments are expected at this time.

The Landscape Characterization Coordination Group has spearheaded the establishment and continuing development of the new interagency Multi-Resolution Land Characteristics Consortium (partners include EPA-EMAP, the GAP Analysis Program, the National Water Quality Assessment Program [NAWQA], the CoastWatch Change Analysis Program, the North American Landscape Characterization Project [includes the National Aeronautics and Space Administration among others], and the EROS Data Center) to provide Thematic Mapper land cover data for the United States. Cooperatively, this group acquired complete Thematic Mapper coverage of the continental United States. Working with this multiagency group, we plan to produce a national land cover map by the end of 1997.

Appendix A   *113*

Analysis of EMAP Data, Particularly Estuaries, and Publication of Results

To date, the Program has published approximately sixty-three peer-reviewed journal articles, and eighteen book chapters; and, it has full papers in twelve scientific conference proceedings. We have set more stringent criteria for measuring the accomplishments of our Technical Directors and Technical Coordinators including higher expectations for publication of the Program's findings in the peer-reviewed, open literature.

The publication productivity of the EMAP-Forest Group was of particular concern to the Committee. In response, the Forest Group has published nine recent journal articles and has an additional ten articles to be published in 1995.

All of the EMAP's Resource and Coordination Groups have refereed publications as an element of our annual evaluation of their progress.

The Program has instituted a Document Tracking System available to all Program participants (but, in particular, to the Program's Executive and Steering Committees) via video text. The Document Tracking System is updated (the current schedule is biannual) to assure better oversight of the quality of vehicles carrying EMAP's scientific and assessment results.

The Program has required the Estuaries Resource Group to curtail broadscale monitoring activities and to focus attention on the evaluation of existing data as suggested by the Committee. Indicator testing in the field is still being conducted but not on a regional grid scale.

Information Management

A peer reviewed, revised Strategic Information Management Plan for EMAP was published in August, 1994.

The "Proof-of-Concept" for the EMAP Data Management System now is completed; and, we have begun to install the operational system for use by EMAP Resource Groups (Estuaries, Forests, and Surface Waters). We should have these all installed by Fall 1994. It will be, however, approximately six months before the complete data sets are available readily to all in the Program as it will take some time for the Resource Groups to populate the new system and to convert their current operations over to the new, uniform system.

We have initiated cooperative development efforts with the EPA's Office of Information Management and the National Biological Survey. We are hopeful that these interactions will ensure relatively transparent data transfers in the future. No final agreements, however, have been reached.

EMAP-Estuaries, Surface Waters and Forests are changing their approach to information management from a Non-Relational Data Structure to an Agency Standard Relational Database Management Structure using Oracle. Within the next six months, this modification will be complete.

Other resource groups who do not have large field data sets will begin using Oracle now. Also, we are focusing on the interface between our data, ArcInfo, and various statistical analysis packages. These interfaces are partially developed but are not expected to be fully operational for some months.

Plans are progressing to make all EMAP data available on the Internet. Tests, to date, have proven successful, only the availability of the validated data sets is slowing the progress in this area.

EMAP-Estuaries is developing with NOAA a compatible and efficient data interface tool that will permit wide access to EMAP-Estuaries data and will promote the use of EMAP-Estuaries data in academic and applied research as well as in State and Federal decision making. This interface will be in addition to the Internet interface (as it is expected that all States, for example, will not have necessary Internet options at this time).

The NRC recommended that the Forest Group, in particular, should develop a comprehensive Information Management Plan. They have not done so, to date. We will be asking all of the Program's groups to do the same now that the "Program System" is ready to be tested fully. They will follow the overall, published Information Management Strategy.

Assessment, Integration and Coordination

The Science Advisory Board (SAB) completed its review of the Assessment Framework for EMAP in late 1993 and suggested it be printed for wide distribution—now done.

We have established EMAP-Center in North Carolina where we plan to develop a critical mass of in-house and university scientists to focus on integration and assessment of multiple resource data.

A competitively-awarded University Partnership Research Agreement with a consortium of universities will be awarded within the next two months to provide additional intellectual leadership in this area. The Partnership Agreement is a novel funding mechanism by which we and the consortium design together a Five-Year Research Plan with the option to renew the agreement for an additional five years, given positive peer reviews of work being done.

We do still have the real difficulty of a limited number of in-house positions to fill all of the functions we believe to be important. This problem is beyond the immediate control of the Program, as you know.

We have appointed a Senior Scientist, Rick A. Linthurst, Ph.D., as EMAP-Center Director. As you may recall, Rick was one of the originators of the EMAP concept.

We have appointed a Chief of Integration and Assessment, Marjorie M. Holland, Ph.D., for the centralized management and coordination of EMAP Technical Coordinators through EMAP-Center. This has improved greatly the coordination of many elements of the Program; nonetheless, much still remains to be done.

Appendix A

EMAP Center has enlisted David Mouat, Ph.D., as a Visiting Scientist to develop a credible and reliable assessment strategy for EMAP. The draft of this document for internal review is expected next month.

The Center has initiated planning for the Mid-Atlantic Integrated Assessment (MAIA). This project, to begin in 1995, will examine key design, indicator, integration, and assessment issues. Two studies of particular interest to the NRC will focus on the robustness of the current design at different scales and with various ecological boundaries, and an appraisal of the benefits to be derived from sampling all resources within randomly-chosen watersheds. Currently, as you know, there is in EMAP no obvious relational structure to the individual resource measurement points. Also, studies on the relationships between land use or cover, landscape metrics and resource condition will be a primary area of investigation. We believe the Mid-Atlantic effort will be an important field research activity to benefit all of EMAP. We will work with other agencies, States, our Regional Offices, and the university community to complete these studies. The university work will be awarded competitively.

Cause-and-Effect Relationships Need More Focused Planning

With the more clearly defined indicator strategy, emphasis on the conceptual models, and clarification of the objectives, I believe we are making progress in this most important area. Over the past six months, EMAP has actively participated in and provided leadership for the development of a new "Integrated Ecosystem Protection Research Program" in the EPA's Office of Research and Development (ORD). This research program links EMAP directly with new as well as on-going cause-and-effect research activities in ORD. This final, peer-reviewed research strategy will be completed in FY 1995 for implementation in FY 1996. Investigator-initiated cause-and-effect research will play a significant role in this program.

Program Management

While it has been made clear that the Program continues to experience a substantial staffing shortfall, considerable progress has been made in the conceptualization and subsequent development of the EMAP-Center, at Research Triangle Park, North Carolina.

A vigorous Director, Rick A. Linthurst, Ph.D, one of the initiators of the Program, has been named.

An individual with established credentials within the ecological research and policy communities, Marjorie M. Holland, Ph.D., has been hired as the permanent Chief for Integration and Assessment to provide strong, scientifically-defensible leadership to the Program's Coordination Groups.

At the Program's headquarters offices, an individual certified as a Senior Ecologist by the Ecological Society of America, Sidney Draggan, Ph.D., has been appointed to provide staff there, and in the Ecological Resource and Coordination Groups, with leadership in development of credible science goals and outputs, enhancement of Program policies and procedures (including adoption of a more traditional peer-review process), and better attention to and oversight of development of performance mechanisms by the Program.

In addition, at the headquarters offices, an individual with a strong background in the ecological sciences, H. Kay Austin, Ph.D., has been enlisted to serve as the Program's permanent Indicators Coordinator as well as its Agency Program Office Liaison (this is, in fact, a superior opportunity since the Agency's Program Offices have considerable interests in indicators; to date, this interest has not been articulated adequately or related to the Program's goals and actual capabilities.)

While we believe these steps are moving us in the right direction, we are faced, as yet, with a limitation in needed positions, a problem that is beyond our control. Whenever possible, we are attempting to fill these voids with visiting scientists.

Involvement of the Scientific Community and Peer Review

As noted earlier, we have instituted an investigator-initiated indicator development research grants program in EMAP. This was administered by the Agency's Office of Exploratory Research through an extensive peer-review process. There were nine meritorious proposals related to ecological condition indicator research that have been funded.

We have increased involvement of the academic statistics and monitoring research communities in the development of EMAP's statistical design through ten nationally-competed Cooperative Research Agreements, totaling $0.80M in FY 1994.

In FY 1994, the Program disseminated a Request for Proposals for studies to be located within the NSF's network of Long-Term Ecological Research (LTER) and Land-Margin Ecosystems Research (LMER) sites.

The Program solicited proposals for the establishment of a University Partnership Research Agreement to strengthen the Integration, Coordination and Assessment activities of the EMAP-Center. Thirteen proposals were subjected to a traditional peer review process that included independent, anonymous mail reviews as well as panel deliberation by a group of experts in the ecological and ecological risk assessment sciences.

We have established peer review policy on a program-wide basis. All of EMAP's Ecological Resource Groups have external, independent peer panels that afford review and guidance to the Program. The Program's past and current associations with the Estuarine Research Federation and the American Statistical Association will stand as the models for external peer-review used by the Program's other Ecological Resource Groups, and Coordination Groups.

Appendix A

EMAP Center has awarded four universities with two training grants each to fund graduate students to participate in various parts of the Program. They are expected to do thesis research on topics of mutual interest.

The Mid-Atlantic studies will be undertaken with university scientists through competitive awards.

Significant increases in the grants program within EPA, including research necessary to advance EMAP concepts, is planned for 1995 and beyond.

Inter-Agency and Intra-Agency Cooperation

EMAP initiated and funded peer-reviewed, Regional EMAP (R-EMAP) projects in all EPA Regions. Two of these projects were summarized at our last meeting with the Committee. Also, full-time appointment of a Program Office Liaison, H. Kay Austin, Ph.D. (from the Agency's Office of Pesticide Programs) and an Interagency Liaison, James K. Andreasen, Ph.D. (from the U.S. Fish and Wildlife Service) has occurred.

Presently, we have established more than twenty-five Memoranda of Understanding and Interagency Agreements with nineteen Federal agencies to support cooperative studies and implementation activities.

Over the past year, the Program has strengthened its interaction with its current Federal agency partners and has established new relationships with others (for example, a Memorandum of Understanding has been entered into with the National Biological Survey and EMAP is developing new relationships with the USGS's NAWQA). Also, existing Interagency Agreements are being revised to establish the focus of research-based activities with the EPA and implementation-based activities with our partner resource management agencies. The EPA component of EMAP has developed an interagency coordinating group for Statistical Design with NAWQA, the Natural Resources Inventory, the Forest Health Monitoring Program, the National Wetlands Inventory, the Bureau of Land Management, and the National Biological Survey; and, it has increased its participation in the Interagency Task Force on Monitoring Water Quality.

EMAP-Estuaries is aggressively pursuing the involvement of other Federal agencies (NOAA and NBS), State resource agencies, and national or regional estuarine programs (Gulf of Mexico Program [GOMP], Chesapeake Bay Program [CBP], and National Estuary Programs [NEPs]). To date, five of the twenty-two coastal states have agreed to modify their monitoring programs to incorporate aspects of EMAP, six NEPs have adopted the EMAP approach in toto or in part, and all three large regional programs (CBP, GOMP, and the Puget Sound Water Quality Authority) have adopted or are adopting aspects of the EMAP approach in their monitoring programs.

Again, we have increased greatly our participation in the Inter-Governmental Task Force on Monitoring Water Quality.

Finally, we have increased our support of the EPA Program Office's monitoring needs in indicator development and development of monitoring designs.

Appendix B

May 4, 1994 letter from Gary J. Foley,
EPA Acting Assistant Administrator
for Research and Development

UNITED STATES ENVIRONMENTAL PROTECTION AGENCY
WASHINGTON, D.C. 20460

MAY - 4 1994

OFFICE OF
RESEARCH AND DEVELOPMENT

Ms. Sheila D. David, Study Director
Committee to Review the EPA's Environmental
   Monitoring and Assessment Program
Water Sciences and Technology Board
National Research Council
2101 Constitution Avenue, N.W.
Washington, DC 20418

Dear Sheila;

I want to thank you and the Committee for your recent review of the Forests and Estuaries components of EMAP. I am pleased that the Committee continues to support the goals of the program and sees progress in those components.

The recommendations in the Estuaries and Forest chapters are both thoughtful and helpful, and I understand that EMAP already has begun to take action on many of them. Matt Bills and I will make sure that all of the recommendations have been given careful attention before making substantial commitments to further implementing either component. I fully agree that articulation of a coherent, consistent, and comprehensive strategic plan is a priority for EMAP, and Ed Martinko is making this a top priority.

As we discussed during our visit, I am concerned that in the executive summary, the Committee questions the degree to which EMAP can meet its goals in a timely and effective manner. It is not clear to me whether this concern has more to do with the technical adequacy of the indicators and sampling design, or the costs and management of the program. Perhaps it would be helpful for me to respond to the Committee from a policy perspective.

Regarding the overall design of the program, I would like to speak to the EPA's reason for investing in EMAP. A critical component of Administrator Browner's National Goals Program is to establish measurable goals for improving the quality of the environment and to monitor progress in achieving these goals. EPA also was advised by its Science Advisory Board, in its report, Reducing Risk, to focus its limited resources on opportunities that offered the greatest potential for risk reduction. When in the past EPA has attempted to determine the quantitative extent of any particular environmental problem so as to assess relative risk, or to track progress resulting from regulatory programs, data from individual intensive

Appendix B

study sites selected for research or management attention seldom have provided the necessary, relevant information.

This is so because most environmental harm results from the aggregation of local actions (e.g., pollution discharges and habitat loss and alteration) on a matrix of differentially susceptible ecological systems. The resulting patterns of harm are spatially complex, making it very difficult to extrapolate from one research site to many, so as to gain an unbiased estimate of the conditions of the whole. On the other hand, the statistical sampling approach developed to assess the status and trends in surface water acidification from acid rain (and upon which the EMAP design was based) provided scientifically sound information that was exactly suited to EPA's decision-making needs.

The issue is not whether intensive monitoring will better meet EPA's needs (it won't), but whether the EMAP approach will. The questions that I hope the Committee will answer in this regard are: (1) do the chosen indicators accurately distinguish between systems about which we should be concerned and those about which we should not; and (2) will the sampling design detect differences in the relative proportions of these situations with sufficient precision to identify with adequate confidence which resources are at greatest risk, and in time to conduct cause/effect studies and institute management changes before significant or irreversible harm occurs to the resource as a whole? What is adequate, in either case, and is the "perfect" the enemy of the "good"?

There are both science and policy issues operating here, and I sense from the panel's comments that it is time to bring these two audiences together. I would be glad to help facilitate such a dialogue if the Committee would find it useful.

The committee continues to express concern over the way EMAP treats cause and effect. Early on, it was decided to focus on indicators of biological condition, rather than indicators of exposure to pollutants or habitat modification because of the uncertainty surrounding the cumulative impact of aggravating or mitigating effects of multiple stresses on ecosystems. It has been a "null hypothesis" in EMAP that collecting exposure and habitat indicator data to be used in an "epidemiological" search for statistical associations between indicators of poor condition and exposure to pollutants or altered habitat would be cost-effective, given the high fixed costs of visiting an EMAP sampling site. I suspect that exposure and habitat data are necessary in evaluating the credibility of the biological condition indicators early in the program, but its cost-effectiveness on a permanent basis needs to be rigorously evaluated.

Borrowing from Koch's Postulates, in no case would epidemiology be sufficient for presumption of a cause-effect relationship. Instead, it should help to prioritize ecosystems for exposure assessment and ecological effects research. ORD is now in the process of taking the long-awaited second step in its integrated ecological research strategy, which is to refocus its ecological research into conducting intensive ecological research at a number of sites and regions across the country, in what has been termed the "place-centered" approach to integrated ecosystems management. It is here that cause-effect research will be centered, and we also expect to test the EMAP design for establishing baseline conditions and trends in ecological indicators at less than national scales. Many EPA Regional Offices are finding EMAP-developed technology useful at finer scales in the REMAP program. This government-wide initiative is coordinated through the White House Committee on Environment and Natural Resources.

Regarding costs, considering the enormous off-budget costs of over-protecting or under-protecting the environment based on inadequate information about environmental status and trends, EPA believes that a scientifically sound EMAP program would be a good investment even at $50-100 Million per year. Under the current discretionary domestic budget constraints, however, this may not be possible. We currently are reviewing our options for further development of the program, and accurate estimates of the cost of implementation are critical. It has been my experience, however, that because of economies of scale in procurement, eventual unit costs for field sampling programs such as EMAP are very difficult to predict with confidence in the early stages.

With regard to program management, having spent most of my career managing complex, multi-organizational programs, I can say with certainty that they are never managed as well as we would like them to be. In the case of EMAP, it has been made especially difficult by severe limitations on hiring, changes in contracts and grants management within EPA, and unique personnel issues beyond ORD's control. If the Committee has knowledge of particularly successful complex programs from which we could learn, I would appreciate specific management recommendations.

Finally, I would like to address the issue of communication. I can appreciate the committee's frustration with the "voluminous literature" of EMAP and the ponderous process of peer-reviewed publication of research results. While this process seems to be speeding up (as of the end of 1993 EMAP counts 244 peer-reviewed technical products), the process of analyzing samples, quality assuring the data, interpretation of results, and peer review seems to take at least a year. Implementation of a permanent program eventually will speed up the process for annual data reports, but that is cold comfort to the Committee.

If there is anything I can do to insure that the Committee has the latest possible information to guide its deliberations in the eleventh hour, I am at your service. In return, I would ask that the Committee be as specific as possible in its critiques and recommendations, for this will allow me to ensure that we are taking appropriate action to address its concerns. As we discussed at our meeting, I would like to offer the services of Jay Messer to you and the Committee to facilitate communication with EMAP, Matt Bills, and myself. Although Jay has not been involved directly with EMAP since his rotational assignment to the United States

Senate in 1991, as you know his knowledge of the program and the issues surrounding its instigation is extensive. He will give this assignment the highest priority. He can be reached as follows:

Dr. Jay J. Messer, Acting Director
Atmospheric Research and Exposure Assessment Laboratory
U.S. EPA (MD-75)
Research Triangle Park, NC 27711
Voice: 919-541-2107
Fax: 919-541-7588
E-mail: messer.jay@epamail.epa.gov

Again, please convey my appreciation to the Committee for their efforts on our behalf. EPA Administrator Browner is committed to insuring that EPA's science is of the highest quality and credibility, and the Committee's careful evaluation of EMAP is paramount to achieving this end.

Sincerely

Gary J. Foley,
Acting Assistant Administrator
for Research and Development

cc H. Matthew Bills
    E. Martinko
    J. Messer

EXECUTIVE SUMMARY AND OVERVIEW

NRC Comment: Some 18 months after the initial report many of the questions raised in the committee's interim report have not yet been answered.

EMAP Response: We agree. However, following the June 1992 report from the NRC, EMAP began to implement many of the suggestions contained in that report and initiated research efforts to answer questions that were raised. In addition, the Assistant Administrator for ORD convened a special workshop, attended by both SAB and NRC members, that dealt with several issues raised in that interim report. Research addressing some of these issues requires a longer period of time to produce results. Questions that require management and organizational answers extend beyond EMAP and are being actively pursued. Thus, EMAP has made significant progress since the interim report although some issues have not been fully addressed.

NRC Comment: The committee foresees many technical difficulties in detecting meaningful trends at scales relevant to policy decisions. This apparent lack of commitment to evaluating the best approach for temporal trend detection is a serious flaw in EMAP program development.

EMAP Response: This comment is pervasive in many comments of the committee. We concur that the description of "meaningful trends at scales relevant to policy decisions" is an area of ongoing discussion that requires resolution prior to implementation of long-term EMAP monitoring. However, our initial technical information and policy discussions are confirming that our approach to detecting changes is developing useful information that is not available elsewhere.

The committee has recommended intensive site-based sampling approaches for the detection of trends. While EMAP concurs with the need for information based on such an approach, EMAP recommends that site-based intensive monitoring information continue to be collected through other ORD, federal, and academic programs, because site-based information alone is insufficient for detecting "meaningful trends at scales relevant to policy decisions".

Concerns on the technical aspects of detecting trends are addressed both here and in the estuaries and forest sections. EMAP has sponsored statistical research on the properties of statistical sampling designs and their ability to detect trends and estimate status. These studies clearly show that detection of trends under EMAP sampling design alternatives is technically feasible. Based on this research and information gathered by the resource groups at least three groups Forests, Estuaries and Surface Waters will be able to meet

DRAFT

Appendix B

the EMAP Data Quality Objectives (DQOs) for estimating status and detecting trends. While the other resource groups have not completed this analysis, initial indicators are favorable. EMAP is continuing discussions with EMAP information users and policy analysts to establish that the DQOs represent meaningful trends, acceptable levels of uncertainty, and appropriate scales.

NRC Comment: EMAP has chosen to use standard federal regions established by Office of Management and budget in its documentation. .... These regions are not standard between federal agencies ..... In addition, they do not capture known geographic, climatological, or ecological regimes or processes.

EMAP Response: EMAP has established Data Quality Objectives (DQOs) on regional areas approximating the size of the standard federal regions. EMAP anticipates that the information will routinely be reported for a variety of the geopolitical regions (standard federal as well as those used by our partner agencies). For EPA information users, for example, we have committed to provide assessments of the status and trends in condition of the ecological resources within the EPA regions to EPA's Regional Administrators. The program is being designed, however, with additional flexibility to respond to a range of scales and regions, dependent upon the question and the DQOs of the information users. While the design phase is proceeding on the basis of the standard federal regions, EMAP fully anticipates that additional regional definitions based on geographic, climatological and ecological regimes and processes will be used in the analysis and assessment phases. The EMAP design is being develop to accommodate such analyses and provide information about the associated uncertainty.

NRC Comment: The committee expresses concern over the information management within EMAP. More specifically there are five significant areas of concern;
 1) No analysis of user requirements...
 2) No system design nor specific information concerning what the system will do...
 3) No short or long term plan to implement IM.
 4) General consensus among the committee that Oracle and Arc INFO are inadequate to handle complex

EMAP Response: Recently, EMAP has provided the NRC panel with a revised strategic plan for information management, along with the results of a peer review panel. In addition, a current status briefing was conducted for members of the NRC panel that included a demonstration of the current information management system. These briefings and documents address these areas of concern; however, we welcome the NRC's additional specific questions and recommendations for changes and will implement those that are within EPA policies and procedures.
 Information management is indeed a complex issue: rapid advances are being made in scientific data management, computer information networks

data from EMAP.
  5) An information system plan should include the transition from short term to long-term. ... and EMAP should use expertise in spatial data processing and handling at other federal agencies.

and hardware; but these must be integrated within EPA's and EMAP's organizational structure. The use of ARC INFO and Oracle as the basis for EMAP's data base management system, was delayed by EPA's procurement of Oracle, but it is currently required as the Agency standard. This assures maximum transferability of information to other EPA as well as federal agency programs. EMAP is represented on the Federal Geographic Data Committee (FGDC) and does interact with expertise through out the federal government. In spite of the fact some federal agencies lag behind the latest scientific developments in information management systems.

NRC Comment: The EMAP documentation is a large and sometimes unclear body of literature and "typical of many large programs" it has lagged behind its initial schedule.

EMAP Response: We concur. The initial schedule was optimistic and the development and growth of the program research results, budget, infrastructure and management has proceeded at a slower rate. In part, the developmental nature of the program, while documented in over 200 reports, scientific papers and plans, has contributed to difficulties in the panel's obtaining a clear picture of EMAP. Some of the lack of clarity in the documentation arises from the diverse set of disciplines working on EMAP and that different people use the same term to mean different things. We have published a bibliography of common terms used within EMAP and have taken action to assure a more consistent set of terminology in all documents.

NRC Comment: Can EPA achieve overall purpose and basic goals in a timely and cost-effective manner?

EMAP Response: The Office of Research and Development has been established with a mission to conduct research including that necessary to implement a comprehensive monitoring and assessment program like EMAP. As stated by the Science Advisory Board, EPA should be conducting a program like EMAP that requires a commitment by EPA and all other federal partner agencies in order to be successful. EMAP's initial success with other agencies and EPA, leads us to anticipate that we will be successful in obtaining additional commitments of staff and budgets. With respect to timeliness issues, EMAP believes that, in an operational mode in partnership with other agencies, results can be delivered in a timely cost-effective manner.
  EMAP agrees with the need to provide additional information on the cost and cost-effective-

Appendix B

	ness of the program. We will continue to develop ways to further reduce costs and increase our effectiveness.
NRC Comment: The primary difference between this statement of objectives and earlier ones is that the analysis of cause-effect relationships has been downplayed. Instead, EMAP now states (perhaps more realistically) that it will "seek associations" between stress and indicators of ecological condition.	*EMAP Response:* The wording of the third objective was changed as the result of a joint meeting between members of the NRC committee, the SAB, and EMAP to clarify the original intent. Previously the objective read "Monitor indicators of pollutant exposure and seek associations between human induced stressors and ecological condition...." We listened to the joint committee's advice and agreed that we should place our highest priority on biological indicators of condition and not the stressors themselves. We believe that we will be able to provide useful information much as an epidemiologist does when they draw associations between national health statistics and possible environmental contaminants. Our preliminary feedback leads us to conclude that the EMAP information will significantly increase the scientific foundation and data available to administrators and policy makers. The SAB guidance, the degree of interagency activities, the reactions by EPA Program Offices, EPA Regions and several States, all point out that EMAP is providing useful information, that is not available elsewhere. EMAP anticipates continuing these interactions and improving the program's ability to provide useful information for policy makers.
NRC Comment: A coherent, consistent, comprehensive strategic plan for EMAP, should be an objective of the highest echelons of EPA that are involved with EMAP.	*EMAP Response:* We agree. We recognize this is long over due and the EMAP executive committee will give high priority to the preparation of this document.
NRC Comment: There is concern about the way "EMAP addresses cause and effect relationships". "It appears to the committee that the question of causality needs more focused planning by EMAP officials."	*EMAP Response:* EMAP has stated that its monitoring effort can not establish cause and effect; however, it does state an emphasis on seeking associations which do contribute to the scientific weight of evidence process for establishing cause and effect. Existing information on causality from the literature and ongoing scientific research is key in the development of conceptual models and indicator selection. Clearly cause and effect research results will play a critical role in this process and in the assessment of ecological conditions. ORD will focus additional efforts on the best strate-

	gies for incorporating cause and effect results into both research and monitoring programs.
NRC Comment: There is a concern with the NRC committee about overall management and coordination within EMAP.	*EMAP Response:* We concur that management and coordination of the program can be improved. While EPA and ORD have made significant commitments of funds to the program, the program is understaffed. The hiring of some full-time staff at EMAP center has resulted in recent improvements in the coordination of the program. However, additional support in both areas is needed.
NRC Comment: EMAP should make better use of sampling and monitoring programs already developed by other agencies and parts of EPA.	*EMAP Response:* The NRC notes that: "The committee is encouraged by the extent to which the EMAP program has been pursuing opportunities for interagency cooperation. Such cooperation can reduce the cost and significantly increase the usefulness of monitoring programs. It will also result in the different government agencies basing their natural resource programs on consistent information." We concur that use of ongoing sampling and monitoring programs and cooperative efforts are critical to EMAP's success as a regional and national monitoring program. EMAP will continue to actively pursue these efforts with both interagency partners and within EPA.

SECTION 2 ESTUARIES

NRC Comment: The committee agrees with the review panel of the Estuarine Research Federation, which doubts that the indices generated by EMAP will have the power to detect the amount of environmental change expected.	*EMAP Response:* No data set presently exists that represents the type of population for which EMAP-Estuaries is attempting to ascertain trends (i.e., large biogeographic regions). Temporal trends related to estuary-specific data sets will not provide all of the information needed to address this issue, i.e. whether they support trends greater than or less than 2% per year for a decade. The type of data being collected by EMAP-Estuaries is the type of data set necessary to investigate the "expected" level of trajectory for any indicator.

At present, EMAP-Estuaries indicators have the potential to detect between a 1-2% change per year for a decade. This is based on an analysis of the Virginian and Louisianian Provinces data sets that will be published in 1994. With the possible exception of some data sets from Chesapeake Bay (which comprises 60% of the Virginian Province), no available historic data can be used to quantitatively estimate the expected long-term trends for biogeographical provinces. This is precisely what |

Appendix B

makes the EMAP-Estuaries data unique and important.

For example, the EMAP design standard is the ability to detect a 20 percent change occurring over a decade. The published information on changes in various indicators shows, however, that some changes occur in estuaries at a much slower rate than this (Stanley, 1993).

The EMAP Data Quality Objective for trends is to detect a +/- 2% per year for a decade. Stanley (1993) documented trends of nutrients, chlorophyll, and bottom dissolved oxygen in the Pamlico River Estuary for 12-24 years at 3 sites (upper, middle, and lower river). Although the representativeness of these three sites for making trends statements concerning the Pamlico River Estuary can be debated, the site-specific data (not regional) presented tends to support a 2% per year change rate. While some information is missing, the overall monthly data at the three sites is available for trend analysis.

Seventy-eight percent of the nutrient-site combinations showed significant trends (7 of 9 sites). Of these seven, six showed trend rates ranging from +/-2.2 to 7.7% per year. The remaining site showed a trend of 1.6% per year. Another stressor indicator (bottom dissolved oxygen) showed only a 0.2% per year change. However, chlorophyll a showed a 6.6% per year change in concentration at a single site in the upper estuary. These rates relate to changes in the measured concentration of an indicator at a site, not the area characterized by a specific condition.

NRC Comment: It seems that in some cases EPA personnel have not researched the published literature.	EMAP Response: EMAP-Estuaries and its cooperators (University of Rhode Island, Rutgers University, University of Maryland, Virginia Institute of Marine Sciences, University of North Carolina, South Carolina Marine Resources Research Institute, Florida Marine Research Institute, Gulf Coast Research Laboratory, Texas A&M University, University of California at Berkeley, and the Southern California Coastal Waters Research Project) have carefully studied the published literature as evidenced by the hundreds of references to this literature in the EMAP-Estuaries publications. EMAP-Estuaries welcomes any additional references and the specific information that should be applied, that the committee would like to suggest.

NRC Comment: By contrast, as is pointed out in Chapter 3, changes in ecosystems can be quite sudden and catastrophic, perhaps too fast to be adequately captured by EMAP's sampling scheme.	*EMAP Response:* If significant change is occurring too fast to be adequately captured by the EMAP-Estuaries design, then we would assume that detecting a 2% per year change would be more than adequate. If these catastrophic changes are ephemeral (i.e., recovery is evidenced in less than the one year between EMAP-Estuaries samples), then we should rethink our definition of catastrophic. Nonetheless, it is not an EMAP objective to detect short term episodic events.
NRC Comment: There is a temptation to think that the next challenge is to carry out pilot projects on new provinces, one after another. However, as pointed out by the Estuarine Research Federation review committee, the real challenge is in obtaining the best possible set of indicators of ecological condition.	*EMAP Response:* EMAP-Estuaries does not disagree; however, we believe that the challenge is to develop a set of ecological indicators that is applicable for all biogeographic provinces. This challenge can only be met by evaluating, the utility, accuracy, precision, and interpretability of our selected indicators in all biogeographic provinces. Our present EMAP-Estuaries strategy calls for the collection of 4 years of data from the Virginian and Louisianian Provinces (the two largest in terms of estuarine resources), followed by intensive evaluation of that data and specific indicator testing and modification in the remaining 10 provinces. Making important decisions, such as the final selection of core indicators, without some testing in the remaining provinces would be imprudent at best and potentially disastrous. We must recognize that the estuarine resources of the Nation are diverse and the indicators must be shown to apply to all of them, prior to implementation of long term monitoring.
NRC Comment: The committee considers the probability-based sampling approach to be satisfactory and flexible enough to meet the problem of how to sample large estuaries, small estuaries, and rivers. However, the value of the resulting data could be significantly enhanced by the addition of methods for detecting step functions of change, including threshold effects. This is an appropriate area for further research.	*EMAP Response:* EMAP-Estuaries, the EMAP Design & Statistics Team, and university cooperators are examining the effect of multiple design modifications on the detection of different types of trends functions, including step functions, nonlinear functions, and various other functions. We agree with NRC that this represents a fertile area for further research.
NRC Comment: The committee is concerned that the proposed prob-	*EMAP Response:* EMAP-Estuaries has, through a joint project with EMAP-Design and Statistics and

Appendix B

ability-based sampling is not adequate for meaningful trend analysis. Consideration should be given to development of an alternative regional-sampling scheme that includes detailed site-based sampling, which would lend itself to more effective trend-analysis procedures.

Oregon State University and several contracted statisticians, conducted analyses using the EMAP-Estuaries data sets for the Virginian and Louisianian provinces to evaluate our ability to discern 2% per year decadal trends. Trends of 1-2% per year for a decade were detectable for the Virginian and Louisianian Provinces from virtually all indicators comprising the EMAP-Estuaries data sets. Only data sets containing fewer than 40-50 observations over the four-year period could not attain the level of trend detection.

While EMAP-Estuaries agrees that the use of site-based intensive sampling is necessary to further evaluate the causes of the observed trends, it is not necessarily required for their detection. In addition, while this approach maybe more effective at detecting trends at a specific site, it has not been shown that it is sufficient for detecting trends within a resource class at province scales. EMAP is cooperating with NOAA's Status and Trends monitoring program, which includes more detailed site-based sampling.

NRC Comment: A more explicit conceptual model is needed to direct indicator research and provide integration across different estuarine provinces.

EMAP Response: EMAP agrees with NRC that the current conceptual model can be improved, and is working to provide a more explicit conceptual model.. This conceptual model will appear in a manuscript, detailing its development, relationships to environmental values and indicators, and relationship to short- and long-term indicator research, to be submitted for publication by Environmental Monitoring and Assessment in 1994.

NRC Comment: Fish sampling should be reexamined carefully and either expanded significantly or dropped. In particular, the measurement of the contaminant body burden as currently implemented is fundamentally flawed because no account was taken of temporal (seasonal variability), and more importantly no account was taken of the relation between body burden and the size (age) of the fish.

EMAP Response: The present fish program concentrates on community measures of the fish species susceptible to the sampling gears used, pathologies, biomarkers, and body burdens. The number of trawls and their lengths have already been increased in EMAP-Estuaries in 1992-1994. The purpose and intent of the EMAP-Estuaries sampling is to reflect conditions during an index period, i.e. a short-temporal window (4-6 weeks) annually. While there was no direct adjustment of body burdens for the size (age) of the fish analyzed, there was care taken in both the Virginian and Louisianian Provinces to composite fish of the same size ranges at each station and across stations. Of course, this approach will not always be successful based on the age structure of the fish populations at each site. However, length data were collected for all fish that were analyzed and EMAP-Estuaries will in its four-year assessments of the Virginian and Louisianian Provinces determine the effects of variability in length (age) on the body burdens determined for a sample. EMAP agrees that this research and analysis should be completed before additional decisions about the future of the fish program are made, i.e. expand or drop.

NRC Comment: The evaluation of current indicators should continue. The determination of the variability of the dissolved oxygen measurements carried out by EMAP-Estuaries is a good example of the care that must be taken in evaluating even a relatively simple indicator.	*EMAP Response:* EMAP agrees with the NRC in this regard. Evaluation of the variability associated with all EMAP-Estuaries indicators is one of the primary steps in the EMAP indicator development strategy. EMAP has and will continue its efforts to fully evaluate the utility of all indicators proposed for use. To this end, it is also important to test specific indicators under a wide range of conditions, including geographic and habitat differences. This necessitates the further testing of indicators in regions outside the Virginian and Louisianian Provinces and could not be accomplished if EMAP were to stop all pilot activities.
NRC Comment: Whenever possible, EMAP-Estuaries should develop remotely-sensed indicators. Candidate measures would include surface chlorophyll and extent of submerged aquatic vegetation.	*EMAP Response:* EMAP agrees with the NRC and has been utilizing remote sensing (aerial photography) to assess the extent of submerged aquatic beds through an interagency project with Fish and Wildlife Service (now National Biological Survey, NBS) and NOAA. More sophisticated remote measures based on satellite imagery are also being evaluated although they appear to have some difficulties based on interference factors. EMAP-Estuaries presently is measuring surface chlorophyll to evaluate its utility as an indicator, primarily to estimate its short-term temporal variability within the index sampling period.
NRC Comment: EMAP-Estuaries should develop indicators of ecosystem function, such as productivity.	*EMAP Response:* EMAP agrees with the NRC, although we have identified few short-term measures of ecosystem function. Since 1991, EMAP-Estuaries has been evaluating several functional indicators of ecological condition. Foremost among these indicators are indices of estuarine trophic state based upon productivity, nutrient concentrations, chlorophyll a, and stable isotopes.

Appendix B

NRC Comment: The benthic index should continue to be examined, particularly the validity of the use of weighting factors in the discriminant analysis.	*EMAP Response:* The benthic index has been and will continue to be examined. Changes in the specific structure of the benthic indices developed for the Virginian and Louisianian Provinces have been made since June, 1993. Validation studies are continuing. In addition, EMAP-Estuaries and the Chesapeake Bay Program (CBP) have been cooperating on a comparative study of the Virginian Province Index and the CBP's Restoration Goals Index (an index without weighting factors). That comparison shows approximately 85-88% overlap with most differences resulting from differences of opinion among benthic ecologists regarding the functional status of some species (i.e., whether species A is an equilibrium or opportunistic species).
NRC Comment: Research must continue on the best way to make measurements of fish pathology and on the best way to separate the various causes of the pathology.	*EMAP Response:* EMAP agrees. EMAP-Estuaries is continuing this research activity at EPA's Gulf Breeze and Narragansett laboratories and through an interagency agreement with the Fish and Wildlife Service (now National Biological Survey (NBS)). The EPA laboratory studies are examining the use of specific pathological indicators as indicators of fish community degradation. These activities focus on lesions and biomarkers, particularly splenic macrophage aggregates. The NBS work uses rearing studies to focus on the linkage between contaminant exposures and the subsequent development of increased density of size of macrophage aggregates culminating in lesions.
NRC Comment: A detailed analysis of the successes and failures of the various pilot and demonstration projects should be undertaken and the best ideas should be incorporated into the indicator plan for EMAP-Estuaries.	*EMAP Response:* EMAP agrees and will establish indicator workshops, including distinguished members of the estuarine research community, to accomplish precisely this goal. These workshops will establish through the results of the pilots, demonstrations, and indicator testing studies a suite of core indicators that will be used in developing a national implementation plan for coastal monitoring.
NRC Comment: Investigator-initiated research should continue to be supported to validate indicators in use and develop new ones.	*EMAP Response:* Largely in response to previous recommendations from the NRC and SAB, EMAP initiated in 1993 an investigator-initiated indicator research program through the Office of Exploratory Research to accomplish this goal. Nine research projects were begun in 1993. Currently, EMAP has solicited and is in the process of reviewing several

	investigator-initiated research proposal for research to be conducted at Long Term Ecological Research (LTER) and Land Margin Ecological Research (LMER) sites.
NRC Comment: EMAP-Estuaries should use the detailed historical data sets already identified in the Virginian province to provide temporal and spatial information on indicators and their ability to detect change.	*EMAP Response:* EMAP-Estuaries has examined several large-scale data sets in the Virginian and Louisianian provinces that have proven useful in determining annual variability. However, EMAP most detailed data sets have proven to be less useful because they; have not used consistent or comparable indicators or methods, their data quality is not known, or they are specific to one estuarine system and not likely be representative of the larger province or regional scale.
NRC Comment: EMAP-Estuaries should give priority to determining the power of these indices to detect the changes that are reasonably expected over the next decade.	*EMAP Response:* We concur. The determination of the power of the presently used indicators to detect trends is a priority issue associated with the four-year assessments of data in the Virginian and Louisianian Provinces.
NRC Comment: EMAP-Estuaries should change to a standard relational database management tool for storage and retrieval in the near future.	*EMAP Response:* EMAP-Estuaries is presently converting from its initial data base storage tool of SAS to an ORACLE-based data base, the current EPA standard. This data base conversion is expected to be completed by mid-1994.
NRC Comment: Alternative multivariate analysis techniques should be carefully explored in addition to the linear discriminant function that was applied to data for the classification of estuarine habitats. EMAP might also explore methods of statistical computing concerning visualization of multidimensional data.	*EMAP Response:* EMAP agrees with the NRC and is currently examining the approaches mentioned. A new interagency agreement with the National Biological Survey is examining the use of GIS-analysis and evaluation as a research tool for EMAP-Estuaries data.
NRC Comment: Compatible data sets and efficient interfaces must be developed between EMAP-Estuaries and other segments of EMAP, NOAA, and other federal agencies with large-scale monitoring information. This would allow EMAP-Estuaries to be used as a model for development of data acquisition and management in other EMAP resource groups.	*EMAP Response:* EMAP agrees with the NRC. EMAP-Estuaries is currently working with NOAA and NBS to develop a retrieval system for EMAP-Estuaries data (a first step to the above goal) that can be accessed and used by other agencies and other public users. EMAP is also participating on the Intergovernmental Task Force on Monitoring Water Quality (ITFM) and the group addressing data management and information sharing. EMAP's Information Management Team is tackling the broader issue of integrating data sets across EMAP Resource Groups.

Appendix B

NRC Comment: A mechanism needs to be developed to ensure thoughtful, detailed analyses and interpretation of data. Outside scientists and managers should be involved at all stages, including review.

EMAP Response: EMAP includes outside scientists and managers in the analysis of all data but particularly for the 4-year cycle data. The Virginian Province Assessment Team is comprised of EMAP personnel, academic scientists, and Agency managers. Their task is to produce the 4-year assessment report for the Virginian Province.

NRC Comment: An updated plan needs to be prepared as to how the reports and analyses will incorporate the data on stressors. There has been a change from the original plans, and the entire EMAP now appears to be reducing its emphasis on stressors.

EMAP Response: EMAP-Estuaries continues to collect pertinent data on stressors to be used in condition or status reviews as well as to assess potential associations among biological indicators and stressors. EMAP, as a whole, is addressing the role of stressors in future monitoring and research activities. Dependent on the results of that effort, EMAP-Estuaries will prepare a plan for the use of internally and externally monitored stressor information within the EMAP context.

NRC Comment: It is imperative to assess EMAP-Estuaries in detail before going on or before adding additional provinces. This assessment should begin as soon as the Virginian Province demonstration completes its four-year cycle.

EMAP Response: EMAP-Estuaries is making significant efforts to analyze fully the four-year cycle data sets for the Virginian Province (1990-1993) and the Louisianian Province (1991-1994) while testing the applicability of indicators in other regions of the country in short, concise pilot studies. It is on the basis of these analyses that EMAP-Estuaries will propose a national implementation design plan and indicator suite. EMAP-Estuaries did stop collecting data in the Virginian Province in 1993 after a full 4-year cycle of data was collected to permit complete analysis of the information. Since the objectives of the EMAP program require comparisons among biogeographic regions, EMAP must develop a set of ecological indicators that are applicable for all biogeographical provinces. Consequently, evaluation of selected indicators in other provinces will be continued while detailed analysis of the Virginian Provinces four-year cycle demonstration is being completed.

NRC Comment: EMAP-Estuaries has begun development of indicators that should be useful to managers of estuarine resources. It now must begin to develop a program for continual involvement of these managers in improving the indicators and reporting and interpreting results.

EMAP Response: EMAP agrees with NRC and, to this end, we have included Agency and State Managers in our workshop teams for the development of the national design and indicator suite for EMAP-Estuaries. In addition, EMAP-Estuaries data is made available for Regional and State use and our interaction with these entities is continual. These efforts are anticipated to continue on a frequent basis.

NRC Comment: The scientific applications of the EMAP-Estuaries data are not yet clear. The present-day open availability of raw EMAP-Estuaries data must be maintained in order to foster the scientific use of the information.	*EMAP Response:* EMAP agrees with the NRC. To date, EMAP-Estuaries data has been transmitted to over 300 users. However, we believe the present system for request and transmittal of data is often cumbersome. Therefore, EMAP-Estuaries is developing, with the Information Management Group within EMAP, a public access system that would allow ready access by modem to all the composite EMAP data. Raw data, inclusive of QA and QC data, will remain in its present form of accessibility.
NRC Comment: Greater emphasis must be placed on the timely production of annual statistical summaries and reports on demonstration projects. Availability of an up-to-date assessment of how EMAP-Estuaries is doing and what it is finding is important to managers and to the scientific community.	*EMAP Response:* EMAP agrees with the NRC and has taken steps in 1993-1994 to reduce the amount of time required before information can be released.
NRC Comment: The quality of the scientific review panel organized by the Estuarine Research Federation was high. EMAP-Estuaries needs to establish a regular, working, review panel of similar quality for each component of the program.	*EMAP Response:* The ERF panel was established as a result of a request from EMAP-Estuaries. EMAP-Estuaries will also establish, in a collaborative effort with NOAA's NS&T Program, a joint review panel to begin work in 1995. We intend to work with this panel to determine the national design and indicator suite for coastal monitoring. In addition, EMAP-Estuaries instituted in 1991, province-specific review panels comprised of state and regional environmental managers and academic estuarine researchers to provide a province-level review of demonstration plans. This combined national review and province review adds depth to the review process and we believe makes EMAP-Estuaries a better program.
NRC Comment: Coordination between EMAP-Estuaries and state, regional, and national monitoring programs should continue to be aggressively pursued. These connections may be the most useful outcome of the program.	*EMAP Response:* We agree that coordination and interaction at the state, regional, and local levels in very important to the success of EMAP. EMAP-Estuaries has worked with six of the 22 coastal states (at their request) to modify their state-level monitoring programs to incorporate all or parts of EMAP's monitoring technology. EMAP-Estuaries continues to work with the Chesapeake Bay (CBP) and Gulf of Mexico (GOMP) Programs to develop their comprehensive monitoring plans. EMAP-Estuaries has worked with seven National Estuary

Appendix B   137

> Programs in the development of their Comprehensive Management Plans and three of these NEPs (thus far) have instituted EMAP-like monitoring programs. The collaborative efforts in California have become the catalyst for the promotion of a combined regional, estuary-specific, and local monitoring program with over 100 participants.

SECTION 3 FORESTS

The following response to the NRC comments was prepared jointly by the Forest Service Forest Health Monitoring Program and EMAP.

NRC Comment: However, it should be emphasized that many of the good features of this program derived from the previously established USFS FHM program and not from EMAP.	*FS/EMAP:* Both agencies view the forest monitoring portion of EMAP as an outstanding example of interagency cooperation. While many of the good features of the program came from FHM, many others came about as the result of the cooperative nature that has developed. Specifically EMAP has helped to improve the overall program by improving the sampling frame, looking at the forest as an ecological resource and not just a forest resource, indicator development criteria, quality assurance, information management structure and the assessment approach. We do not view the contributions from each agency separately, but rather as one federal government program.
NRC Comment: EMAP Forests relies too heavily on a purely "epidemiological model." "Such models have little utility in predicting how nutrient cycles, nutrient losses, or biodiversity of ecosystems change in response to stress." It was recommended that EMAP-Forests develop a "theoretical basis from which predictions can be made of general types of forest response to different types of stress."	*FS/EMAP Response:* We agree that one must have a theoretical rationale which takes into account how forests respond to different types of stressors. However, the objective of EMAP is to describe the ecological condition, detect changes, and draw associations between condition known stressors. EMAP will rely on research from EPA ORD, the Forest Service, other federal agencies and the academic community to develop models to "predict response to stress". Consequently, Forest Health Monitoring looks at a wide range of parameters in assessing forest health status and trends. These parameters include factors as diverse as tree crown condition, lichen communities, and soil chemistry. Our overall monitoring rationale is as follows: We can identify a small number of key indicators that reflect the overall health of the forest ecosystem. Using cost-efficient probabilistic sampling, we can use these key indicators to estimate status (or

NRC Comment: EMAP-Forests should choose a set of indicators as soon as possible and then conduct the staff work necessary to establish sampling methods and convey these to field crews.

a beginning point). We can then, by sampling over time, use these indicators to detect broad changes in the forest ecosystem.

FS/EMAP Response: We concur with this recommendation and are concentrating efforts in this area. Several indicators have been evaluated through EMAP's indicator development process and have reached operational status. A number of others are in the process of being evaluated. We recognize the need to add indicators "as soon as possible", particularly in the areas of biotic integrity and aesthetics. However, we weigh this urgency against the need to make sure each indicator receives sufficient and thorough scientific scrutiny during the development process.

Sampling methods are written and provided to the field crews during extensive training sessions each spring.

NRC Comment: The committee recommends that the current design of four-year plot rotations be replaced or augmented by a design in which some plots are revisited each year.

FS/EMAP Response: EMAP-Forests has conducted and continues to conduct research on alternative sampling designs over time and space. EMAP has sponsored statistical research at Oregon State University to investigate the properties of statistical sampling designs and their ability to detect trends and estimate status. The research shows that the FHM four panel sample approach would have increased power with an additional panel of sites with annual revisits, especially in the early years of the program. EMAP-Forests continues to investigate the benefits and costs associated with specific alternatives on the number of sites to be revisited annually and whether these sites should be annually revisited forever or new annual revisit sites should be started in future years. We are concerned about the impact on the individual site from continuous annual revisits. We plan to make a decision prior to implementing long-term monitoring, concerning supplementing our sampling with subsets of sites with annual revisits.

NRC Comment: EMAP-Forests should develop a comprehensive information-management plan that outlines user requirements, examines long-term implementation, and fits in with the overall plan for the information-management system.

FS/EMAP Response: We agree with this recommendation. Although still early in the development stage, Forest Health Monitoring has already collected a wide array of data. Our database has already grown to the point that it cannot be managed by the scientists who gathered the data. The existing SAS data base has been adequate for the initial development and testing of

Appendix B

indicators, but will prove inadequate in the long run. We have been working interactively with EMAP's Information Management Coordinator in developing the overall EMAP Information Management system that will be developed prior to implementation. Our information and data management, like our science, must be performed by a professional trained in that field. The Forest Health Monitoring database development effort is now being led by a professional database manager. We are also in the process of the transitioning from SAS data management software to an Oracle/ARC INFO system which is the EPA agency standard.

NRC Comment: The results of the Forest Health Monitoring efforts be published in peer-reviewed science journals.	*FS/EMAP Response:* We agree. Peer-reviewed publications are and will continue to be a factor in the performance evaluations of our technical staff. Forest Health Monitoring scientists continue to communicate FHM program information to the broader scientific community by publishing in recognized, peer-reviewed journals. At least seven articles have been published in such journals as Environmental Monitoring and Assessment, The Canadian Journal of Forest Research, Communications in Soil Science and Plant Analysis, and the International Journal of Climatology. In addition, work by FHM was included in at least eight symposia proceedings and book chapters.
NRC Comment: The Forest Health Monitoring "should not be fully implemented until the results of demonstration projects have been fully evaluated and realistic estimates of cost to EPA and other agencies is available.	*FS/EMAP Response:* Forest Health Monitoring is being developed in an incremental fashion. States are gradually being brought into the program and new indicators are gradually being evaluated and, if validated, adopted. With this approach, the program will not be fully implemented without a detailed accounting of the cost of various activities and the total program. For example, in the indicator development phase, cost and time constraints are an important determinant in whether an indicator can be selected for use. We are constantly aware that Forest Health Monitoring can afford to spend only so much time on each plot and so much money for analysis. Therefore, when comparing one indicator with another, the information value of each indicator or measurement is evaluated relative to its cost.

140	*Appendix B*

NRC Comment: The committee made several references in the body of the report concerning approaches other than monitoring based on a probabilistic sample. For example, the committee seems to feel that a more appropriate approach to Forest Health Monitoring might be long-term research at key locations coupled with modeling. | *NRC Comment:* The committee made several references in the body of the report concerning approaches other than monitoring based on a probabilistic sample. For example, the committee seems to feel that a more appropriate approach to Forest Health Monitoring might be long-term research at key locations coupled with modeling.

NRC Comment: The committee made several references in the body of the report concerning approaches other than monitoring based on a probabilistic sample. For example, the committee seems to feel that a more appropriate approach to Forest Health Monitoring might be long-term research at key locations coupled with modeling.	*FS/EMAP Response:* We agree that the quoted sentence probably cannot be defended because of the unqualified "only." This last sentence by the Committee is an excellent one; they make an important point. Forest Health Monitoring must keep in mind that change, especially on the local scale, can sometimes be sudden. Nevertheless, EMAP-FHM was not designed to detect site specific changes and other programs are capable of identifying catastrophic events.
NRC Comment: This may be due to the fact that the original purpose of the USFS FHM program was to detect the effects and extent of diseases and insects.	*FS/EMAP Response:* The original purpose of the USFS FHM program was to determine the status and condition of the forest ecosystem of which insects and diseases play significant roles.
NRC Comment:. To date, the indicators proposed by EMAP-Forests clearly address productivity only.	*FS/EMAP Response:* Although productivity, in its broadest definition, is an important part of EMAP-FHM, biodiversity, sustainability, aesthetics, and extent are of equal importance.
NRC Comment: There is apparently no single set of indicators that is to be sampled nationwide.	*FS/EMAP Response:* Differences in the sets of indicators in the demonstration studies are the direct result of the indicator development process. All of the indicators used in the Detection Monitoring Demonstration are measured on all plots across all forest types, ecosystems and states (16 in 1994).
NRC Comment: The Western Pilot Study (EPA, 1992b) showed that trying to make many measurements on each plot is logically infeasible.	*FS/EMAP Response:* We have operationally measured as many as 15 indicators per plot within a one day visit. We have done this in different pilots and demonstrations across the U.S.
NRC Comment: No structure or criteria for selecting indicators have (sic) been accepted by both USFS and EPA.	*FS/EMAP Response:* The USFS and EPA came to an agreement on the criteria and process for selecting indicators about two years ago. This process has allowed the addition of lichen community structure, plant biodiversity, wildlife

habitat, photosynthetic active radiation and ozone bio-indicator plant measurements to the Detection Monitoring Demonstration being conducted in 16 states in 1994.

Appendix C

EMAP Documents Reviewed by NRC Committee

1. The EMAP Design Perspective - September 30, 1990.

2. EMAP Landscape Characterization Research and Implementation Plan - May 1990.

3. Indicator Development Strategy for EMAP - December 1990.

4. Design Report for EMAP Part I, May 14, 1990.

5. Near Coastal Program Plan for 1990: Estuaries - November 1990.

6. A Review of the U.S.G.S. National Water Quality Assessment (NAWQA)-Pilot Program by National Research Council.

7. The EMAP Design Perspective - presentation by Scott Overton, Oregon State University, March 26, 1991.

8. The EMAP Landscape Characterization 1991 Pilot Project Summaries, March 1991.

9. National Research Council Review of Environmental Monitoring and Assessment Program. EPA presentation March 26-27, 1991.

10. Guide to EMAP Documents for NRC Review by EPA.

11. Summary Update for the EMAP Landscape Characterization Research and Implementation Plan - by Douglas J. Norton, May 1991.

12. List of EMAP material titles and status of drafts from EPA.

13. List of EMAP Interagency Contacts, program components, and telephone numbers, April 30, 1991.

14. Review of the Near Coastal Report dated March 20, 1991 by a panel convened by the Estuarine Research Federation.

15. May 31, 1991, letter to Rick Linthurst regarding the panel results of the ERF review of EPA's EMAP-NC, conducted April 16-18, 1990.

16. Review of the EMAP Near Coastal Program Plan for 1990 by the Estuarine Research Federation, April 16-18, 1990.

17. EMAP-NC Responses to ERF Comments.

18. Peer review comments by James Hornbeck on Monitoring and Research Strategy for Forests - Questions and Answers.

19. Binkley's preliminary comments on EMAP documents in response to memo from K. Bruce Jones dated February 14, 1991.

20. Comments about EMAP forests by Joseph B. Yavitt.

21. February 28, 1991, letter to Steve Paulsen from Kenneth L. Dickson, Final Report of EMAP Surface Waters Component Peer Review Panel - February 1991.

Appendix C   *145*

22. February 7, 1991, memo to EMAP Surface Waters Component Review Panel from Kenneth L. Dickson, Final Draft of the Surface Waters Component Peer Review Panel.

23. February 18, 1991, memo to Roger Blair from Dr. Steve Paulsen and Dr. David Larsen, EMAP Surface Waters Reconciliation Memo.

24. Peer Review of EMAP Wetlands to Environmental Research Laboratory Office of Research and Development, EPA, Corvallis, Oregon by Peer Review Panel, November 28-30, 1990.

25. Reconciliation letter dated January 25, 1990 to Dr. Roger Blair from Nancy Leibowitz, "Research Plan for Monitoring Wetland Ecosystems" - Reviewers' concerns and suggestions.

26. Review of the EMAP-Arid Strategic Plan - March 8, 1991, by a technical review team.

27. Report of the Ecological Monitoring Subcommittee of The Ecological Processes and Effects Committee - Evaluation of the Indicators Report for EMAP - August 2, 1990.

28. Comments on Ecological Indicator Report for EMAP by Glenn W. Suter, II, Environmental Sciences Division, ORNL.

29. February 1, 1991, memo to EMAP ADs, TDs, TCs, and a few others from Don Charles, Acting Technical Coordinator for Indicators, "The Indicator Development Strategy for the Environmental Monitoring and Assessment Program" (cover letter).

30. January 22, 1991, memo to Charles (Mel) Knapp from Bob Hughes, Review of the EMAP Indicator Development Strategy.

31. January 14, 1991, letter to Charles (Mel) Knapp from David Rapport, Research Coordinator on his review of the document, "The Indicator Development Strategy for EMAP".

32. January 21, 1991, letter to Donald Charles, EPA from Brock Bernstein, Ph.D. on his review of the Indicator Development Strategy.

33. January 23, 1991, letter to Charles (Mel) Knapp from Richard Latimer, Acting Technical Director, EMAP-NC. Comments on the "Indicator Development Strategy" document from Dr. John Scott (SAIC) and Dr. Dan Campbell.

34. December 24, 1990, memo to Don Charles from Ann Fairbrother on her comments on EMAP Indicator Strategy.

35. May 15, 1990, letter to Dr. Rick Linthurst, Director from Linda Young, Chair, American Statistical Association, et al., on their concerns before reviewing the statistical progress of EMAP. April 1990.

36. December 20, 1989, memo to Rick Linthurst from ASA Review Committee for EMAP on Grid Design for Characterization and Tier 1 Sampling.

37. January 25, 1991, letter to Mr. Douglas Norton from Gene Thorley on the final version of the report entitled, Report of the Peer Review Panel: "EMAP Landscape Characterization Research and Implementation Plan", Environmental Monitoring and Assessment Program - June 1990.

38. EMAP - Responses to National Research Council Questions (booklet) July 1991.

39. Characterizing Dissolved Oxygen Conditions in Estuarine Environments by S. Weisberg (Versar, Inc.), J. Summers (EPA),

Appendix C 147

A. Holland (Versar, Inc.), J. Kou (Versar), V. Engle (Technical Resources, Inc.), D. Breitburg (Academy of Natural Sciences), R. Diaz (Virginia Institute of Marine Science).

40. Appendix to Question 12 from NRC committee concerning Memorandums of Understanding and Interagency Agreements.

41. Appendix to Question 15 from NRC committee concerning Peer Review Comments and Reconciliations.

42. Comments on Ecological Indicator Report for the Environmental Monitoring and Assessment Program - Glenn W. Sutter II - Environmental Sciences Division, ORNL. March 14, 1990.

43. June 4, 1991, memo to Thomas E. Dixon, Acting Associate Director, EMAP from William Laxton, Director of Technical Support Division, comments on EMAP Ecological Indicator Report.

44. February 26, 1991, memo to Robert Blair, Chief, Watershed Branch from Donald Charles. Reconciliation of comments on "The Indicator Development Strategy for the EMAP Program".

45. Reconciliation Memorandum EMAP, Arid Ecosystems Strategic Monitoring Plan, 1991 - Summary of Peer Review Comments.

46. Report of the Ecological Monitoring Subcommittee of the Ecological Processes and Effects Committee, August 2, 1990, Evaluation of the Indicators Report for EMAP - A Science Advisory Board Report.

47. Final Report of the Surface Waters Component Peer Review Panel - dated February 1991.

48. Appendix 4-B. The Interpenetrating Design for EMAP by W. Scott Overton, April 1990.

49. EMAP Responses to Comments by the Science Advisory Board, Ecological Processes and Effects Committee, Ecological Monitoring Subcommittee on "Evaluation of the Indicators Report for EMAP" - March 1991.

50. EMAP - Plan for Converting the NAPAP Aquatic Effects Long Term Monitoring (LTM) Project to the Temporally Integrated Monitoring of Ecosystems (TIME) Project - Internal Report.

51. Appendix 4-A to Question 4: Changes and Trends, by Scott Overton, May 1991.

52. Summary, Results of the First Sampling Design Workshop for Estuaries, May 9-10, 1991, Columbia, MD - May 21, 1991.

53. EMAP - Example Environmental Assessment Report for Estuaries - May 1991.

54. Implementing EMAP data management facilities - Progress Review, January 17, 1990.

55. Science Advisory Board's Review of the EMAP plan. Letter dated July 30, 1991 to The Honorable William Reilly from Raymond Loehr, Chairman, Science Advisory Board and Kenneth Dickson, Ecological Processes and Effects Committee. Report to the Ecological Monitoring Subcommittee of the Ecological Processes and Effects Committee - Evaluation of the Program Plan for EMAP, July 1991.

56. Letter dated July 30, 1991 to The Honorable William Reilly from Dr. Raymond Loehr, Dr. Kenneth Dickson, and Dr. Richard A. Kimerle, Chairman, Marine Disease/Diagnostic Task Group. Report of the Marine Disease Diagnostic Task Group of the Ecological Processes and Effects Committee - Evaluation of the Proposed Center for Marine and Estuarine Disease Research, July 1991.

Appendix C

Letter dated March 21, 1991 to Dr. Robert Menzer, Director, Gulf Breeze Environmental Research Laboratory from Kenneth L. Dickson, Richard A. Kimerle, and Edward S. Bender. Response to two questions addressed as part of the SAB review.

57. Letter dated August 11, 1991 to Sheila David from K. Bruce Jones, Associate Director, Terrestrial Ecosystems (EPA), with Data Confidentiality Report by Dr. Sue Franson. Copies of reports dealing with information management, a draft copy of a user mission needs assessment and information management plans for Near Coastal and Forest field activities, and a manuscript highlighting 1991 indicator development for EMAP-Forests.

57a. Proposed Policy and Rationale: Use of Data Collected Under the Auspices of EMAP - July 1991.

57b. Graph - Complete Software Life Cycle.

57c. Graph - Exhibit 1-3 System Category EEI Matrix.

57d. EEI-1-Mission Needs Statement - EMAP - July 1991.

57e. Appendix A. EEI-1 Mission Needs Statement EMAP - Agroecosystems.

57f. Appendix D. EEI-1 Mission Needs Statement EMAP - Information Center.

57g. Appendix E. EEI-1 Mission Needs Statement EMAP - Forests.

57h. Appendix F. EEI-1 Mission Needs Statement EMAP - Great Lakes.

57i. Appendix G. EEI-1 Mission Needs Statement EMAP - Integration & Assessment.

57j. Appendix I. EEI-1 Mission Needs Statement EMAP - Near Coastal.

57k. Appendix K. EEI-1 Mission Needs Statement EMAP - Wetlands.

57l. Binder: FY91 Indicator Evaluation Field Study for Environmental Monitoring and Assessment Program - Forests (EMAP-F) June 1991.

58. Data Management for Near-Coastal Demonstration Project, August 1990.

59. Surface Waters Monitoring and Research Strategy - February 1991.

60. Monitoring and Research Strategy for Forests, March 1991.

61. Agroecosystem Research Plan 1991 - February 1, 1991 - Peer Review.

62. Arid Ecosystems Strategic Monitoring Plan, 1991.

63. EMAP Research Plan for Monitoring Wetland Ecosystems - January 1991.

64. Example Environmental Assessment Report for Estuaries, May 1991.

65. Design Report for EMAP - August 1991.

66. Surface Waters Implementation Plan - Northeast Lakes Pilot Survey, Summer 1991 - June 1991.

67. The Indicator Development Strategy for EMAP, April 1991.

Appendix C

68. Forest Health Monitoring Plot Design and Logistics Study, August 1991.

69. Annual Report: Forest Health Monitoring, New England 1990.

70. Surface Waters Implementation Plan - Northeast Lakes Pilot Survey, May 1991 - REVIEWS.

71. Environmental Monitoring and Assessment Program: Surface Waters Implementation Plan - Northeast Lakes Pilot Survey, June 1991 - Response to Reviews.

72. The Environmental Monitoring and Assessment Program - Responses to National Research Council Questions October 1991, Part 1 and Part 2, October 1991.

73. Monitoring the Condition of Agroecosystems, by Julie Meyer and George Hess, August 4-7, 1991.

74. Memorandum dated March 10, 1992 to William K. Reilly, Chair from Erich Bretthauer, Executive Director of Task Force. 45-Day Task Force on Science Recommendation, Requested 1-Page Action Plans.

75. Review of EMAP Statistics and Design (Review meeting held November 4-6, 1991, San Francisco, CA). Prepared by the American Statistical Association Committee on EMAP, November 1991.

76. Implementation Plan for Monitoring the Estuarine Waters of the Louisianian Province, 1991, October 1990.

77. EMAP Estuaries Component: Louisianian Province 1991 Demonstration Field Activities Report, January 1992.

78. A Selection of Forest Condition Indicators for Monitoring Abstract, January 1990.

79. Booklet: Safeguarding the Future: Credible Science, Credible Decisions, March 1992.

80. Agroecosystem Monitoring and Research Strategy, May 1991

81. Arid Ecosystems Strategic Monitoring Plan, June 1991.

82. EMAP - Estuaries Virginian Province 1990 Demonstration Project Report, January 15, 1992.

83. Estuaries Virginian Province Logistics Plan for 1991, April 1991.

84. Near Coastal Louisianian Province 1992 Sampling, Field Operations Manual, March 1992.

85. Indicator Development Strategy (Updated - Version 1.5) June 2, 1992.

86. Final Peer Review Panel Report of EMAP Great Lakes Monitoring and Research Strategy (Draft EPA Document February 1992). Prepared for U.S. EPA. Submitted by Research and Evaluation Associates, Inc. April 15, 1992.

87. The Relationship Between EMAP and CASTNET, July 1992.

88. EMAP Great Lakes Response to Peer Review Panel Report, June 1992.

89. EMAP-Estuaries Virginian Province 1990 Demonstration Project Report, June 1992.

Appendix C

90. Technical Design Proposal, Clean Air Status and Trends Network (CASTNET), External Review Draft, February 18, 1992.

91. EMAP Estuaries: A Review Organized and Facilitated by the Estuarine Research Federation, March 30-31, 1992. Report dated April 10, 1992.

92. EMAP Estuaries: Draft - 1991 Virginian Province Field Activities Report, December 1991.

93. Executive Summary of the Third Review of EMAP Estuaries, Organized and Coordinated by the Estuarine Research Federation, April 10, 1992.

94. EMAP Great Lakes Monitoring and Research Strategy, June 1992.

95. EMAP - Great Lakes Response to the Peer Review Panel Report, April 1992.

96. EMAP - Program Plan, February 1991.

97. Monitoring and Research Strategy for Forests, March 1992.

98. EMAP-IM 93 Tactical Plan Component 1: Consolidated Statement of Work, October 10, 1992.

99. Information Management Strategic Plan: 1993-1997, Version 1.6, Septemper 30, 1992.

100. FY91 Forest Health Monitoring, Western Pilot Operations Report.

101. EMAP-Estuaries Virginian Province 1990 Demonstration Project Report, June 1992.

102. EMAP-Surface Waters Response to Peer Review Comments, January 1993.

103. Summary of ORD Workshop to Address and Resolve Issues and Concerns Regarding EMAP, Research Triangle Park, North Carolina, September 1992.

104. EMAP Response to the National Research Council's Interim Report, June 1992.

105. EMAP-Surface Waters 1991 Pilot Report, February 1993.

106. Report on the Ecological Risk Assessment Guidelines Strategic Planning Workshop-Risk Assessment Forum, February 1992.

107. "Hexagon Mosaic Maps for Display of Univariate and Bivariate Geographical Data" by Daniel B. Carr, Anthony R. Olsen, and Dennis White. (Abstract from Cartography and Geographic Information Systems, Vol 19, No. 4, 1992, pp. 228-236, 271).

108. Forest Health Monitoring 1991 Statistical Summary.

109. Ambient Water-Quality Monitoring in the United States, First Year Review, Evaluation and Recommendations, December 1992.

110. EMAP-Arid Colorado Plateau Pilot Study - 1992: Implementation Plan, January 1993.

111. EMAP-Agroecosystem 1992 Pilot Plan (April 3, 1992).

112. Annual Statistical Summary: Agroecosystems: A Hypothetical Example, August 1990.

113. EMAP-Response to Congress on the National Research Council June 1992 Report, April 12, 1993.

114. Implementation of a National Monitoring Program, April, 1993.

115. Reprint: Comparing Sampling Designs for Monitoring Ecological Status and Trends: Impact of Temporal Patterns.

116. Statistical Summary: EMAP-Estuaries Louisianian Province - 1991, January 1993.

117. EMAP FY 1994 Issue Planning Paper, March 15, 1993.

118. Summary of ORD Workshop to Address and Resolve Issues and Concerns Regarding EMAP (Research Triangle Park, NC, September 1992), January 7, 1993.

119. Program Guide, May 1993 (handout May 24-25, 1993 meeting). (Updated version - June 1993 in files).

120. Louisianian Province Demonstration Report, EMAP-Estuaries: 1991, dated October 1993.

121. Environmental Monitoring and Assessment Program Master Glossary, October 1993.

122. SAB Review of EMAP's Draft Assessment Framework, September 30, 1993.

123. Statistical Summary: EMAP-Estauries Louisianian Province - 1992, September 1993.

124. Statistical Summary: EMAP-Estuaries Virginian Province - 1991, January 1994.

125. Forest Health Monitoring, 1991 Georgia Indicator Evaluation and Field Study (no date).

126. Forest Health Monitoring: Southeast Loblolly/Shortleaf Pine Demonstration Interim Report, 1994.

127. Manuscript: Regional Scale Monitoring of Indicators of Trophic Condition of Lakes, May 1993.

128. Binder: Multi-Resolution Land Characteristics Consortium, Documentation Notebook, January 1994.

129. A. Forest Health Monitoring, 1992 Annual Statistical Summary. B. Reconciliation of Reviewer Comments from EPA.

130. A. Assessment Framework, February 1994. B. Project Summary: EMAP Assessment Framework. C. Review of the Assessment Framework Draft document by EPA Science Advisory Board and EMAP's responses to their review.

131. EMAP Status Estimation: Statistical Procedures and Algorithms.

132. A. Landscape Monitoring and Assessment Research Plan, 1994. B. Landscape Ecology Review Panel Report. C. Reconciliation of all major reviewers' comments.

133. EMAP Information Management Strategic Plan: 1993-1997. March 1994.

134. EMAP-Surface Waters Stream Pilots in the Mid-Atlantic Highlands, February 1994.

135. Environmental Monitoring and Assessment Program for the Great Lakes, FY92 Status Report.

Appendix C 157

136. Agroecosystem Pilot Field Program Plan - 1993.

137. Chesapeake Bay Watershed Pilot Project, March 1994.

138. Agroecosystem Pilot Field Program Report - 1992, March 1994.

139. A. Indicator Development Strategy, March 1994. B. Indicator Development Strategy Reconciliation Memo, March 31, 1994.

140. R-EMAP, Regional Environmental Monitoring and Assessment Program, September 1993.

141. EMAP-Surface Waters 1994 Streams Pilot Field Operations and Methods Manual, USEPA, Office of Research and Development, EPA/620/R-94/004, March 1994.

142. EMAP-Surface Waters Lakes Field Operations - Volume 1, USEPA, EPA/620/R-93, May 1993.

143. Stream Indicator and Design Workshop, EPA/600/R-93/138 July 1993.

144. A. Entity Relationship Diagram, Forest Health Monitoring, March 1, 1994. B. Case Dictionary Entities and Attributes. Forest Health Monitoring, May 20, 1994.

145. User Interaction and Planning Support for EMAP-IM, Technology Transfer Design Workshop Notes, Feb. 7, 1994.

146. A. Summary of the Proof of Concept Joint Application Design (JAD) Session, September 25, 1992. B. Summary of the Proof of Concept Joint Application Design (JAD) Session II, January 15, 1993.

147. Draft - Building the EMAP Data Set Directory. Prepared by Dr. Jeffrey B. Frithsen and Dr. Donald E. Strebel, Versar, Inc. for Dr. Robert Shepanek, Office of Modeling, Monitoring Systems and Quality Assurance, USEPA, Washington, D.C., February 23, 1994.

148. Letter from Gary Foley, Acting Assistant Administrator for Research and Development concerning the recommendations made in NRC report on EMAP's Forest and Estuaries. May 4, 1994.

Appendix D

Biographical Sketches of Committee Members

RICHARD FISHER, *chair*, serves as professor of forest soils, head of the Forest Science Department, and director of the Institute for Renewable Natural Resources at Texas A&M University. His research interests are in soil–plant interactions in both temperate and tropical systems and the use of experimentally derived knowledge for managing forest productivity. He is a fellow of the Soil Science Society of America and the Society of American Foresters and is president-elect of the National Association of Professional Forestry Schools and Colleges. He served on the National Research Council Committee on Forestry Research and is currently Co-Editor-in-Chief of *Forest Ecology and Management*. Dr. Fisher received his B.S. from the University of Illinois in 1964 and his Ph.D. from Cornell University in 1968.

PATRICK L. BREZONIK received a B.S. in 1963 from Marquette University, an M.S. in 1965 from the University of Wisconsin, and a Ph.D. in water chemistry in 1968 from the University of Wisconsin. Dr. Brezonik is professor of Environmental Engineering and Director of Water Resources Research Center at the University of Minnesota. He was chairman of National Research Council Panel on Nitrates in the Environment in 1975-1978; and a member of the National Research Council Committee on Restoration of Aquatic Ecosystems. He is currently a member of the National Research Council's Water Science and Technology Board. Dr. Brezonik's research interests are eutrophication of lakes, nitrogen dynamics in natural waters, nutrient chemistry, acid rain, trace metals in natural waters, and organic matter in water.

INGRID C. BURKE is a Research Associate, Natural Resource Ecology Laboratory, Colorado State University. She is currently involved in interdisciplinary research programs investigating the control of plant productivity, soil organic matter turnover, and trace gas flux in the Great Plains. Dr. Burke was a member of the EPA's EMAP Landscape Characterization Panel and NSF's Conservation and Restoration Biology Panel, 1990. She has a B.A. in biology from Middlebury College and a Ph.D. in botany from the University of Wyoming. She is a member of the Association of Women in Science, the American Association for the Advancement of Science, American Institute of Biological Sciences, the Ecological Society of America, and the Soil Science Society of America.

LOVEDAY L. CONQUEST received her B.A. in mathematics in 1970 from Pomona College, her M.S. in statistics from Stanford University in 1972, and her Ph.D. in biostatistics from University of Washington in 1975. Dr. Conquest is a biostatistician working in the areas of environmental monitoring (experimental design, sampling design, data analysis/interpretation), natural resource management (e.g., fisheries, forestry, ecology), and related areas. She is associate dean of the College of Ocean and Fishery Sciences and associate professor in the School of Fisheries' Center for Quantitative Science (CQs) in Forestry, Fisheries, and Wildlife at the University of Washington, and heads the Statistical Consulting Laboratory for CQS. She has provided consulting services to other researchers, environmental consulting firms, and public agencies. Dr. Conquest is a Fellow of the American Statistical Association.

THURMAN L. GROVE received a B.A. in 1966 from Wilkes College and a Ph.D. from Cornell University in ecology and soil science in 1982. He is currently Professor of Soil Science, Assistant Dean of the College of Agriculture and Life Sciences, and Director of International Programs at North Carolina State University. His research interests are biogeochemistry, agroecology, and international development.

JOHN E. HOBBIE received a BA in 1957 in ecology from Dartmouth College, an M.A. in 1959 from the University of California, and a Ph.D. in zoology in 1962 from Indiana University. He was Assistant Professor to Professor of zoology at North Carolina State University from 1965-1975. Dr. Hobbie is currently Senior Scientist and Director, Ecosystems Center at the Marine Biology Laboratory, Woods Hole, Massachusetts. Dr. Hobbie's research interests are arctic and antarctic limnology, heterotrophic bacteria in aquatic ecosystems, estuarine ecology, and the global carbon cycle.

TIM KRATZ received a B.S. in 1975 (botany) from the University of Wisconsin-Madison, an M.S. (ecology and behavioral biology) from the University of Minnesota-Twin Cities in 1977, and a Ph.D. (botany) from the University of Wisconsin-Madison in 1981. From 1981-1985 he was Project Associate and Site Manager, Northern Lakes Long Term Ecological Research Project. He is currently Assistant Scientist and Site Manager for that project. His research interests include limnology, landscape/lake interactions, wetland formation, and landscape ecology.

ANNE E. MCELROY received a B.S. in aquatic biology from Brown University in 1976 and a Ph.D. in oceanography from the Massachusetts Institute of Technology/Woods Hole Oceanographic Institution Joint Program in 1985. From 1986-1991 she was an assistant professor in the Environmental Sciences Program at the University of Massachusetts-Boston. Dr. McElroy currently serves as the director of the New York Sea Grant Institute and holds an appointment as an associate professor at the Marine Sciences Research Center at the State University of New York at Stony Brook. Her research interests concern how aquatic organisms interact with toxic chemicals in the environment.

JOHN PASTOR received his B.S. in geology in 1974 from the University of Pennsylvania, an M.S. in soil science in 1977 from the University of Wisconsin-Madison, and a Ph.D. in forestry and soil science in 1980 from the University of Wisconsin-Madison. Dr. Pastor is currently Research Associate, Natural Resources Research Institute at the University of Minnesota; Adjunct Professor, Department of Ecology and Behavioral Biology, University of Minnesota; and Adjunct Professor, Department of Fisheries and Wildlife, also at the University of Minnesota. His research interests are northern ecosystems, nutrient cycling, climate change, forest productivity, timber management, and landscape ecology.

JAMES N. PITTS received a B.S. in 1945 from the University of California, Los Angeles, and a Ph.D. in chemistry from the University of California, Los Angeles in 1949. He has been a member of National Research Council Advisory Board on Military Personnel Supplies for the Committee on Textile Functional Finishing, and Chairman of the Panel of Polycyclic Organic Matter for the Committee on Kinetics of Chemical Reactions. Dr. Pitts is Professor Emeritus in the Department of Chemistry and past Director of the Statewide Air Pollution Research Center at the University of California, Riverside. He was also an Adjunct Professor in the Department of Chemistry and Biochemistry, California State University, Fullerton, and Coordinating Instructor in Air Pollution in the Extension Division of the University of California, Irvine. Dr. Pitts is currently a research chemist at the University of California, Irvine, California. His research interests include fundamental processes in photochemistry and photooxidations and their application to the atmospheric chemistry of photochemical smog, acid rain, airborne toxics, and mutagenic and/or carcinogenic pollutants.

Appendix D

SAUL B. SAILA received a B.S. from University of Rhode Island in 1949, an M.S. in fishery biology in 1950 from Cornell University, and a Ph.D. in fishery biology from Cornell in 1952. He was a fishery biologist at the Rhode Island Department of Agriculture and Conservation. He was professor of oceanography and chief scientist, Office of Marine Programs at University of Rhode Island 1975-1988. In 1988 he became Emeritus Professor and consultant. His research interests are in fish population dynamics and stock assessment.

TERENCE R. SMITH received a Ph.D. in 1971 from Johns Hopkins University. He attended the Graduate School of Management, University of Rochester, New York: Doctoral Program in Applied Economics in 1975-1976. Dr. Smith is currently Chairman, Department of Computer Science, University of California, Santa Barbara, Professor of Computer Science, and Professor of Geography, University of California, Santa Barbara. He is also Associate Director, National Center for Geographical Information and Analysis, Co-Director, Remote Sensing Unit, Co-Director, Center for the Study of Spatial Cognition and Performance, and Associate Director, Community and Organization Research Institute. Dr. Smith's research interests are machine intelligence, spatial databases, spatial cognition, and motion planning.

SUSAN STAFFORD is a forest biometrician at the Department of Forest Science, Oregon State University. Dr. Stafford consults with forest science researchers on the design of experiments in forest ecology; forest genetics; and on the collecting, handling, and analysis of data. She is also data manager for the H.J. Andrews Experimental Forest and director and creator of the Forest Science Data Bank at Oregon State University. Dr. Stafford received her Ph.D. in applied statistics in 1979 from State University of New York, College of Environmental Science and Forestry.

MICHAEL J. WILEY received a B.G.S. in 1973, an M.S. in 1976, and a Ph.D. in 1980 from the University of Michigan. He is Associate Professor, School of Natural Resources, University of Michigan. Dr. Wiley's research interests are ecology of river systems, benthic invertebrates, and fisheries management. Prior experience includes Associate Professional Ecologist, Illinois Natural History Survey, 1984-1987. Dr. Wiley is a member of the Ecological Society of America, American Fisheries Society, and North American Benthological Society.